MOLECULAR
BIOLOGY
INTELLIGENCE
UNIT

Papillomaviruses in Human Cancer:
The Role of E6 and E7 Oncoproteins

Massimo Tommasino

Deutsches Krebsforschungszentrum
Angewandte Tumorvirologie
Heidelberg, Germany

AUSTIN, TEXAS
U.S.A.

Springer-Verlag Berlin Heidelberg GmbH

MOLECULAR BIOLOGY INTELLIGENCE UNIT

Papillomaviruses in Human Cancer:
The Role of E6 and E7 Oncoproteins

LANDES BIOSCIENCE
Austin, Texas, U.S.A.

International Copyright © 1997 **Springer-Verlag Berlin Heidelberg**

Originally published by Springer-Verlag, Heidelberg, Germany in 1997

 Springer

International ISBN 978-1-4757-6129-0

While the authors, editors and publisher believe that drug selection and dosage and the specifications and usage of equipment and devices, as set forth in this book, are in accord with current recommendations and practice at the time of publication, they make no warranty, expressed or implied, with respect to material described in this book. In view of the ongoing research, equipment development, changes in governmental regulations and the rapid accumulation of information relating to the biomedical sciences, the reader is urged to carefully review and evaluate the information provided herein.

Library of Congress Cataloging-in-Publication Data

Papillomaviruses in human cancer: the role of E6 and E7 oncoproteins / [edited by] Massimo Tommasino.
    p. cm. — (Molecular biology intelligence unit)
  Includes bibliographical references and index.
  ISBN 978-1-4757-6129-0   ISBN 978-1-4757-6127-6 (eBook)
  DOI 10.1007/978-1-4757-6127-6
  1. Viral carcinogenesis. 2. Papillomaviruses. 3. Viral proteins. 4. Oncogenic viruses. 5. Oncogenes. I. Tommasino, Massimo, 1958– . II. Series.
    [DNLM: 1. Papillomavirus, Human. 2. Oncogene Proteins, Viral.
3. Papovaviridae Infections--complications. 4. Neoplasms--etiology.
5. Cell Transformation, Neoplastic. 6. Viral Vaccines. QW 165.5.P2 E995 1997]
RC268.57.E15 1997
    616.99'4071—dc21
DNLM/DLC
for Library of Congress
                      97-19205
                      CIP

Publisher's Note

Landes Bioscience produces books in six Intelligence Unit series: *Medical, Molecular Biology, Neuroscience, Tissue Engineering, Biotechnology* and *Environmental*. The authors of our books are acknowledged leaders in their fields. Topics are unique; almost without exception, no similar books exist on these topics.

Our goal is to publish books in important and rapidly changing areas of bioscience for sophisticated researchers and clinicians. To achieve this goal, we have accelerated our publishing program to conform to the fast pace at which information grows in bioscience. Most of our books are published within 90 to 120 days of receipt of the manuscript. We would like to thank our readers for their continuing interest and welcome any comments or suggestions they may have for future books.

Shyamali Ghosh
Publications Director
Landes Bioscience

CONTENTS

1. Human Papillomaviruses and Cancer: A Retrospective 1
   Harald zur Hausen
   1.1. The Origins of Papillomavirus Research 1
   1.2. A Rabbit Papillomavirus System Causing Cancer 2
   1.3. Papillomavirus Infections in Cattle 4
   1.4. Several Human Papillomavirus Types 5

2. Regulation of E6 and E7 Oncogene Transcription 25
   Frank Rösl and Elisabeth Schwarz
   2.1. Introduction 25
   2.2. In Vivo Expression of the E6 and E7 Oncogenes 27
   2.3. Viral and Cellular Factors for E6/E7 Oncogene
        Transcription 29
   2.4. The Consequences of Integration on Viral Gene
        Expression 50
   2.5. The Role of the Genetic Background: Transcriptional
        Regulation of Viral Gene Expression in Nonmalignant
        and Malignant Cells 54

3. E6 Protein 71
   Felix Hoppe-Seyler and Martin Scheffner
   3.1. Introduction 71
   3.2. Expression and Structural Aspects 72
   3.3. Immortalizing and Transforming Activities 73
   3.4. Mechanisms of Immortalization 74
   3.5. Role of E6 in the Viral Life Cycle 89
   3.6. Therapeutic Perspectives 90

4. E7 Protein 103
   Massimo Tommasino and Pidder Jansen-Dürr
   4.1. Introduction 103
   4.2. Domain Structure of the Molecule 103
   4.3. Subcellular Localization 109
   4.4. Immortalizing and Transforming Activities 111
   4.5. Cellular Pathways Targeted by E7 113
   4.6. E7 Protein and Apoptosis 124
   4.7. Conclusion 125

5. Immunological Aspects of the E6 and E7 Oncogenes:
   Tools for Diagnosis and Therapeutic Intervention 137
   Ingrid Jochmus and Lutz Gissmann
   5.1. Introduction 137
   5.2. Humoral Immune Response Against E6 and E7 138
   5.3. Cellular Immune Responses Against E6 and E7 147
   5.4. HPV E6 and E7 as Potential Therapeutic Vaccines 156

Index 167

ABBREVIATIONS

Ad	adenovirus
aa	amino acids
A	alanine
C	cysteine
D	aspartate
E	glutamate
F	phenylalanine
G	glycine
H	histidine
I	isoleucine
K	lysine
L	leucine
M	methionine
N	asparagine
P	proline
Q	glutamine
R	arginine
S	serine
T	threonine
V	valine
W	tryptophan
Y	tyrosine
APC	antigen presenting cell
bp	base pair
β-gal	β-galactosidase
BPV	bovine papillomavirus
CAT	chloramphenicol-acetyl-transferase
CDK	cyclin-dependent-kinase
CDI	cyclin-dependent-kinase inhibitor
C/EPB	CCAAT/enhancer binding protein
CIN	cervical intraepithelial neoplasia
CKII	casein kinase II
CPB	CREB-binding protein
CR	conserved region
CREB	cAMP response element binding protein
cyclic AMP	cyclic adenosine-2',3'-phosphate
cRNA	complementary ribonucleic acid
CRPV	cottontail rabbit papillomavirus
C-terminus	carboxy-terminus
CTL	cytotoxic T lymphocytes
DNA	deoxyribonucleic acid
DTH	delayed type hypersensitivity
E6-AP	E6-associated protein
E6BP	E6-binding protein
E2TA	E2 transactivator protein

ELISA	enzyme linked immunosorbent assay
EV	epidermodysplasia verruciformis
FACS	fluorescence activated cell sorter
GRE	glucocorticoid-responsive element
GST	glutathione-S-transferase
HIV	human immunodeficiency virus
HLA	human leukocyte antigen
HPV	human papillomavirus
HPLC	high performance liquid chromatography
Ig	immunoglobulin
IT-RIPA	in vitro transcription-translation RIPA
kDa	kilodalton
LAMP	lysosomal-associated membrane protein
MCP-1	monocyte chemoattractant protein-1
MHC	major histocompatibility complex
mRNA	messenger RNA
NCR	non-coding region
NF1	nuclear factor 1
N-terminus	amino-terminus
ORF	open reading frame
PALA	N-(phosphoacetyl)-L-aspartate
PBL	peripheral blood lymphocytes
PCNA	proliferating cell nuclear antigen
PEF-1	papillomavirus enhancer binding factor 1
PKC	protein kinase C
PP2A	phosphatase 2A
pRb	retinoblastoma protein
RIPA	radio-immunoprecipitation assay
RNA	ribonucleic acid
RR	relative risk
SDS	sodium dodecyl sulfate
SDS-PAGE	SDS-polyacrylamide gel electrophoresis
SV40	simian virus 40
tAg	SV40 small t antigen
TAg	SV40 large T antigen
TAF	TBP-associated factor
TBP	TATA-binding protein
TEF	transcriptional enhancer factor
TFII	transcription initiation factor
Th 1 or 2	T helper cells type 1 or 2
TNF	tumor necrosis factor
UBF	upstream binding factor
URR	upstream regulatory region
WB	Western blotting
wt	wild type

EDITOR

Massimo Tommasino
Deutsches Krebsforschungszentrum
Angewandte Tumorvirologie
Heidelberg, Germany
Chapter 4

CONTRIBUTORS*

Lutz Gissmann
Chapter 5

Felix Hoppe-Seyler
Chapter 3

Pidder Jansen-Dürr
Chapter 4

Ingrid Jochmus
Chapter 5

Frank Rösl
Chapter 2

Martin Scheffner
Chapter 3

Elisabeth Schwarz
Chapter 2

Harald zur Hausen
Chapter 1

*All contributing authors are affiliated with
Deutsches Krebsforschungszentrum
Angewandte Tumorvirologie
Heidelberg, Germany

PREFACE

In the last decade, hundreds of studies on the human papillomavirus oncoproteins, E6 and E7, have clarified part of their roles in altering fundamental cellular events, such as cell cycle progression, apoptosis and differentiation. In this book, we have summarized several of these studies with the hope of offering to the reader a comprehensive overview on the biology of human papillomaviruses, and in particular on known functions of the two viral oncoproteins. We apologize to those colleagues whose studies have not been specifically cited or whose contributions have not received adequate emphasis. Therefore, we invite the reader to refer to the several excellent review articles mentioned throughout each individual chapter, where many original publications are discussed in detail.

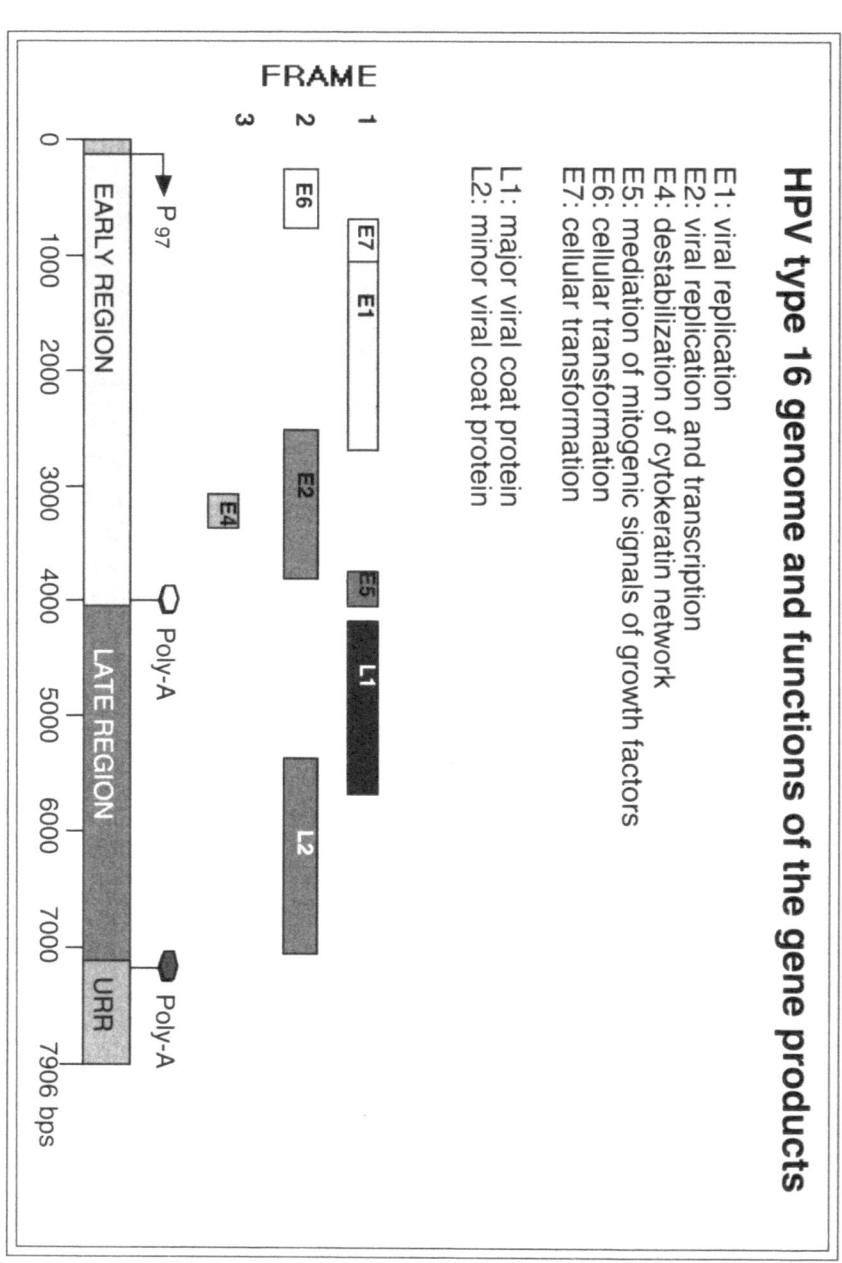

CHAPTER 1

Human Papillomaviruses and Cancer: A Retrospective

Harald zur Hausen*

1.1. THE ORIGINS OF PAPILLOMAVIRUS RESEARCH

The discovery in the second part of the last century of bacteria as causative agents of numerous contagious diseases, including anthrax, diphtheria, typhoid fever and plague, and a widespread disease of high morbidity and mortality—tuberculosis (whose infectious origin had been disputed at that time), created an atmosphere of expectations. This led to the anticipation that even other diseases without epidemiological evidence for an infectious etiology, as for example, cancer and rheumatic fever, may also be caused by infections.

Early attempts in the past decade and the beginning of this century to demonstrate an infectious etiology of warts and papillomas in animals and humans were performed with crude extracts of canine, bovine and human warts. Intraspecies transmission was successful in several of these experiments.

Modified from of a review article: Roots and perspectives of contemporary papillomavirus research. J Cancer Res Clin Oncol 1996; 122:3-13.

Papillomaviruses in Human Cancer: The Role of E6 and E7 Oncoproteins, edited by Massimo Tommasino. © 1997 Landes Bioscience.

Iwanovsky (1894) in Russia and Beijerinck (1898) in Amsterdam demonstrated the filterability of the cause of tobacco mosaic disease and passed the responsible agent through filters known to retain all bacteria identified up until then.[1,2] Their approach stimulated similar attempts with materials obtained from animal and human warts. McFadyan and Hobday (1898) reported the first experiments to transmit warts in dogs by using such cell-free extracts and proved successfully the viral nature of the causative agent.[3] Ciuffo in 1907 demonstrated the cell-free transmission of human warts.[4] This represented a milestone in papillomavirus research. Thus, within the first decade of this century the viral etiology of warts came to be understood; however, 40 years elapsed before the structure of the viral particles was elucidated. After 1907, a number of additional studies revealed the transmissibility of other warts, particularly genital warts and laryngeal papillomatosis (for a review see ref. 5), however, these results provided no more basic understanding of the infectious nature of the wart virus.

1.2. A RABBIT PAPILLOMAVIRUS SYSTEM CAUSING CANCER

Shope in 1933 learned from rabbit hunters in Kansas and Iowa of tumors which originated around head and shoulders, and sometimes formed large horny warts with fissured surfaces, occasionally presented in bizarre formations resembling antlers.[6] He collected such materials and demonstrated their infectious etiology. The inoculation of cell-free extracts resulted in papilloma development in both cottontail and domestic rabbits, and even tolerated heating up to 65°C.[6] Rous and Beard reported one year later the development of malignant tumors when these papillomas were transplanted into muscles or internal organs of the rabbits.[7] The same authors also noted in 1935 the spontaneous occurrence of malignant growth in domestic rabbits after previous exposure to this virus.[8] Syverton and Berry reported the first malignant conversion in the natural host, the cottontail rabbit.[9]

A substantial proportion of the initially benign developing lesions in domestic rabbits, developed within several months into squamous cell carcinomas. This system emerged as the first experimental model for an infectious origin of skin cancers. Rous and colleagues analyzed interactions of this infection with chemical carcinogens and published between 1938 and the early 1950s numerous important papers on syncarcinogenic effects between this infection and tarring of the skin or treatment of the skin with defined chemical carcinogens.[10,11] Systemic application of viral preparations and subsequent tarring of the skin led to virus-containing papillomas at increased frequency, shorter development periods and to a more rapid conversion to malignant growth.

Strauss et al demonstrated with an electron microscope the structure and composition of papillomavirus particles in human warts and Crawford and Crawford elucidated the physical properties of this viral DNA.[12,13] Later on, the rabbit papillomavirus was renamed the cottontail rabbit papillomavirus (CRPV) which has continued to play an important role in papillomavirus research. Ito and Evans demonstrated subsequently that inoculation of viral DNA was sufficient to induce carcinomas in domestic rabbits.[14] They revealed for the first time that a papillomavirus genome may act as a solitary carcinogen. Tumor induction was also achieved by the inoculation of DNA preparations directly from papillomas of the infected rabbits.[14]

In spite of an increasing number of studies on human papillomavirus types and a somewhat diminished attention on CRPV over the past two decades, several studies in this system have contributed to our understanding of papillomavirus biology. In 1992 Han et al reported a role of genetic determinants in the regression and progression of CRPV-induced papillomas.[15] Zeltner and colleagues analyzed differences in the transcription patterns of viral oncogenes in CRPV-induced papillomas of cottontail and domestic rabbits, with increased transcriptional levels in proliferating papilloma cells of the latter.[16] This provided a mechanistic explanation for the increased frequency of carcinoma development in domestic rabbits. In addition, the CRPV system turned out to be important for vaccination studies with nondenatured L1 proteins or virus-like particles composed of the L1 protein.[17,18]

1.3. PAPILLOMAVIRUS INFECTIONS IN CATTLE

Bovine papillomaviruses (BPV) represent another animal papillomavirus system that has been studied intensively in order to understand papillomaviral biology and genome organization. Magelhaes in 1920 reported the first successful transmission of bovine warts by filtrates after inoculation of these filtrates intravenously into calves.[19] Skin wounding subsequently resulted in the development of typical papillomas. Olsen and Cook demonstrated in 1951 that this virus infection is not species-specific.[20] They transmitted BPV to horses and induced sarcoids. Eight years later the same group identified this virus as the cause of urinary bladder tumors in its natural host and also of enzootic bovine hematuria.[21]

The induction of tumors by bovine papillomavirus (BPV) after inoculation into newborn hamsters or mice was demonstrated in the early 1960s.[22,23] The first publication appeared approximately at the same time on transformation of fetal bovine cells in tissue culture by BPV preparations.[24,25] The first report on different types of bovine papillomaviruses appeared shortly after the demonstration of the heterogeneity of the human papillomavirus group.[26]

The analysis of bovine papillomaviruses played an important role in the elucidation of the molecular biology of the papillomavirus group. In 1980 Lowy and colleagues showed that 69% of the BPV genome was required to induce transformation of tissue culture cells.[27] The whole BPV genome was sequenced in 1982, thereby enabling the deduction of the basic genome structure from these data and the characterization of individual open reading frames.[28]

In recent years, interest in the use of bovine viruses as model systems for carcinogenesis has gradually diminished, mainly due to the availability of cancer-linked human papillomavirus types and in vitro test systems to study their biological activity. Besides molecular analyses on genome regulation and DNA replication, studies of syncarcinogenic effects between these virus infections and environmental carcinogens, initiated in 1980 by Jarrett,[29] are ongoing.

1.4. SEVERAL HUMAN PAPILLOMAVIRUS TYPES

The structural identity of papillomaviral particles as revealed electron-microscopically, the failure of in vitro cultivation techniques and the initial absence of serological tests for HPV resulted in the assumption, documented frequently in older dermatological textbooks, that different histological types of warts and condylomata acuminata represent local variants of infections by the same virus. Experiments performed by Almeida and Goffe raised the suspicion that this may not be correct.[30] These authors observed agglutination of viral particles from common warts and condylomata acuminata with sera from wart carriers. By taking sera from condyloma patients, however, only particles from genital warts were agglutinated.[30] This suggested a partial antigenic cross-reactivity of the two virus preparations. Today it is evident that this cross-reactivity was an accidental observation, most likely resulting from antibodies in the sera of common wart carriers reacting specifically with HPV types in genital warts.

One of the first experimental hints for the existence of different HPV types was published in 1974: radioactive cRNA prepared from plantar wart virus DNA hybridized efficiently with plantar warts, but either inefficiently or not at all with DNA from some other warts, and also not with DNA from genital warts.[31,32] Since some of the negative warts used in these experiments contained particles electron-microscopically, this led to the conclusion that their genomic structure must be different and should thus represent new HPV types.

Restriction endonuclease cleavage techniques subsequently permitted the demonstration of the existence of genetic heterogeneity of HPV preparations in 1976.[33] This was followed a year later by two other publications,[34,35] which revealed totally different restriction cleavage maps of new isolates and identified them as new types. In a meeting in Mobile, Alabama in 1978, an agreement between the few groups working on papillomaviruses was reached to use a common nomenclature.[36]

In the following years we would witness an explosion in the identification of new HPV types, which (by including some of the not yet fully characterized new isolates) now well exceeds 100 geno-types (Table 1.1).[37,38] Recent studies on the evolution of HPV point to an ancient history: some nonhuman primate PV types prove to be more closely related to specific human genotypes than several of the latter among each other.[39,40]

The biological significance of the enormous heterogeneity of this virus group is presently not understood. They obviously compete successfully with each other in the human environment, which appears to represent the sole host. This may reflect an adaptation to specific cellular microenvironments. It is very likely that the number of HPV genotypes will increase well beyond its present number. Presently, anogenital and skin types are relatively intensively studied. It remains to be seen whether a similar heterogeneity type also exists in the oral cavity and possibly even in the respiratory or digestive tract.

1.4.1. A rare skin Papillomatosis frequently progressing into skin cancer: epidermodysplasia verruciformis contains specific HPV types

In 1922 Lewandowsky and Lutz described a rare hereditary condition characterized by an extensive verrucosis of confluent flat warts.[41] Also, squamous cell carcinomas were found to develop within pre-existing papillomas preferentially at sun-exposed sites.[42-44] In their initial description the authors did not recognize the infectious component in this condition, as the terminology for describing this condition accurately defines the symptoms. A high level of consanguity in epidermodysplasia verruciformis (EV) families suggested an autosomal recessive mode of inheritance.[45] It took slightly more than 20 years before the viral etiology of the verrucosis was recognized, initially by heteroinoculations of cell-free extracts from the papillomas and the induction of flat warts,[46,47] subsequently by the electron-microscopic detection of papillomaviral particles within the nonmalignant warts.[48-50] Particularly through the contributions of Jablonska and colleagues in Warsaw, this syndrome was increasingly recognized as a model for carcinoma induction by viruses of the papova group.[44]

*Table 1.1. Papillomavirus genotypes and their associated diseases**

HPV Genotypes	Disease
Cutaneous HPV Genotypes	
1	plantar warts
2, 4, 26, 27, 29	common warts
3, 10, 28	flat warts
7	Butchers' warts, oral papillomas in HIV patients
5, 8, 47	benign and malignant EV lesions
9, 12, 14, 15, 17, 19-25, 50	EV lesions
36	actinic keratosis, EV lesions
37	keratoacanthomas
38	melanoma; malignant cutaneous lesions
41, 48	cutaneous squamous cell carcinomas
49	flat warts under immunosuppression
60	epidermoid cysts
63	myrmecia warts
65	pigmented warts
75-77	common warts in renal allograft recipient
78	cutaneous lesions
Mucosal HPV Genotypes	
6, 11	genital warts, laryngeal papillomas
13	oral focal epithelial hyperplasias
30	laryngeal carcinomas; anogenital
32	oral focal epithelial hyperplasias; oral papillomas
16, 18, 31, 33, 35, 39, 45, 51, 52, 56, 58, 59	anogenital intraepithelial neoplasias and cancer
57	oral papillomas and inverted papillomas of the sinuses and nasal cavity; genital warts in children; anogenital intraepithelial neoplasias
66	cervical carcinomas
70	vulvar papillomas
72, 73	oral papillomas (HIV patients); anogenital intraepithelial neoplasias

* For more details see refs. 133, 134.

In 1978 and subsequently, the collaboration of the Warsaw group with Orth and co-workers in Paris turned out to be very fruitful. They showed that carcinomas in epidermodysplasia verruciformis patients contained specific HPV types, most frequently HPV 5, but also other types such as types 8, 14, 17, 20 and 47.[35,51] Unfortunately, up to today no virus-containing cell lines have been established from EV tumors or premalignant lesions. Thus, the system did not permit direct studies on the role of viral transcripts and gene products for the proliferative phenotype of the lesions and for malignant conversion. The regularity of the presence of specific HPV types in skin cancers of EV patients, however, argues strongly in favor of a viral role in the development of these malignancies.

In addition, the system provided the first hint of a cooperative effect between a human papillomavirus infection and an environmental factor (solar exposure) in human carcinogenesis.[42-44]

1.4.2. Papillomaviruses in nonmelanoma skin cancers

Stimulated by the results reported for skin cancers of EV patients, a number of groups initiated studies on skin cancers in other patient groups. Nonmelanoma skin cancers (squamous and basal cell carcinomas) represent the most common malignancy in the Caucasian population worldwide.[52,53] Organ allograft recipients and other patients under long-lasting and severe immunosuppression are particularly "high risk" groups for the development of these cancers, which again most frequently develop at sun-exposed sites. During the past decade a number of reports described the presence of HPV DNA, usually in a small proportion of these cancers.[54-58] In some studies, however, known types of HPV were detected with remarkable frequency.[59-61] Very recently, two groups found remarkably high percentages of HPV positivity in squamous and basal cell carcinomas of these patients, and identified a substantial number of putative new types: Shamanin et al reported HPV DNA in more than 70% of these tumors.[38] These data were confirmed (although with a slight difference in the spectrum of types identified) by Berkhout et al who detected HPV DNA in 81% of the cancers.[62]

Even less frequently, in a limited number of reports the presence of identifiable HPV genomes has been recorded in occasional squamous cell carcinomas of the skin of immunocompetent patients.[63,64] An exceptional situation was revealed in periungual squamous cell carcinomas which almost regularly contained HPV 16 DNA.[65,66] Two recent reports show, however, that squamous and basal cell carcinomas of immunocompetent patients reveal a high degree of HPV positivity (Shamanin et al, submitted for publication and ref. 62). The report by Shamanin et al suggests that more than 50% of these cancers contain HPV DNA, and hitherto unidentified novel HPV sequences.

These reports may open an exciting new field and appear to link one of the most common forms of human cancers to HPV infections. Obviously, the mere demonstration of HPV DNA in a specific form of human cancer does not prove an etiological relationship. It clearly represents, however, a clear-cut precondition for subsequent studies on viral causation and provides the necessary tools for further testing. It appears that our understanding of the relationship of HPV to skin cancer has reached the stage we were in 1983 or 1984 when HPV infections were linked to cancer of the cervix.

1.4.3. Papillomaviruses in cancers of the cervix and the anogenital tract

Since an early publication by Rigoni-Stern in 1842 (which does not reach the standards of contemporary epidemiological approaches) cancer of the cervix has been suspected to have an infectious component in its etiology.[67] I will not try to summarize all the attempts made to link this cancer to sexually transmitted bacterial, protozoal, or herpes viral infections, since excellent reviews cover this subject (for example, see ref. 68).

The hypothesis that cancer of the cervix may be caused by papillomaviruses was put forward at Key Biscayne at a meeting of the American Association for Cancer Research in January 1973. The proceedings of this meeting were published with substantial delay in a supplemental issue of *Cancer Research* in 1976 which also contains a one-page summary of the hypothesis.[69] This hypothesis was based in part on negative results of our group in the analysis of cervical cancer biopsies for herpes simplex virus type 2 DNA, and also in part on a number of anecdotal reports, published over a period of more than 70 years on occasional malignant conversion of genital warts into squamous cell carcinomas, usually in vulval and penile sites (reviewed in ref. 70). In 1974 we published our first report, whereby we attempted to find HPV DNA in cervical cancer and genital wart biopsies, by hybridizing their DNA with cRNA obtained from plantar wart HPV DNA.[29] Since both types of tumors did not react, in spite of the presence of typical HPV particles in some of the condylomata, we wrote in this paper: "The negative result for some of the warts indicates either the scarcity of transformed papilloma cells within these biopsies or their transformation by a different agent."[70] We suspected the existence of several HPV types from this time period on.

Substantial encouragement to study the putative HPV etiology of cervical cancer came from a 1976 publication by Meisels and Fortin who identified koilocytotic atypia of the cervix as the manifestation of a papillomavirus-induced cytopathic change and attempted to separate this condition from "true" dysplasias.[71] The subsequent identification of typical papillomavirus particles in such cells by electron-microscopy underlined these observations.[72]

It took 3-4 more years and the establishment of cloning procedures, before the first genital HPV type—HPV 6—was isolated from genital warts, cloned and characterized.[73,74] This paved the way for the subsequent isolation of HPV 11, a closely related agent, from a laryngeal papilloma and from genital warts[75] and cloning of the still most popular HPV types—HPV 16 and 18—by Dürst and Boshart,[76,77] who were at that time still students in the laboratory. Within a period of several months it also became clear that the typical precursor lesions of anogenital cancer, cervical intraepithelial neoplasias and Bowen's disease also most frequently contained these viruses.[78]

The isolation of genital HPVs directly from cervical cancer biopsies, their immediate availability for the scientific community and the identification of cell lines derived from cancer of the cervix containing HPV genomes resulted in a rapid expansion of the field.[77] Within a relatively short period of time it became clear that specific genes (E6, E7) of those viral types found in cancer of the cervix are transcribed regularly in cancer cells and that most tumor cell lines as well as primary biopsies originating from these cancers harbored the viral DNA in an integrated state.[79] The integrational pattern showed some regularities as far as the opening of the viral ring molecule is concerned, in that the 5'-end of the E2 open reading frame was usually involved, often resulting in deleted 3'-ends of this gene which obviously was not functional in these tumor cells.

Considerations initially derived from other tumorvirus systems, and subsequently from the HPV analyses, tried to explain some of the most difficult aspects in order to understand the up to then putative role of these agents in human carcinogenesis. It became obvious that the latency period between primary infections and carcinoma development commonly lasted several decades, that only a low number of the infected individuals eventually developed the respective form of cancer, that the arising cancers were almost uniformly of monoclonal origin, and that there existed an interesting cooperative effect between some of these infections and chemical or physical carcinogens as outlined in the experiments performed with Shope papillomavirus or in cancer development in EV patients exclusively at sun-exposed sites. In order to reconcile these observations with an assumed causal role of viral infection, an intracellular control of viral oncogene synthesis or function had been proposed.[80-84] It was speculated that under conditions which lead to the failure of this control system viral oncoprotein function is unimpaired and would result in at least some early cellular proliferative changes.

One of the necessary requirements to test this hypothesis was the proof that viral oncogene expression is required for proliferative changes and, if possible, also for the maintenance of the malignant phenotype. After the demonstration by Yasumoto and colleagues in 1986 that HPV type 16 DNA-induced malignant transformation of NIH3T3 cells,[85] two groups simultaneously reported the immortalization of human foreskin keratinocytes by HPV 16 DNA transfection.[86,87]

Subsequently it was shown that only E6 and E7 genes are required for immortalization[88] and that under certain conditions either E6 or E7 genes possess immortalizing properties.[89,90]

One of the key questions whether the expression of these genes is necessary for the growth of cervical carcinoma cells was addressed by von Knebel Doeberitz et al.[91-93] Either by using inducible E6/E7 antisense constructs or by an hormone-inducible switch-off of HPV transcription they demonstrated, in the cervical carcinoma cell lines tested, that expression of at least one of the viral oncogenes is necessary for the proliferative and the malignant phenotype.

Prior to these studies several experiments suggested that, in spite of the requirement of viral oncoproteins for cell proliferation, their mere expression was not sufficient for immortalization or malignant conversion. This became particularly evident from early studies of Pereira-Smith and Smith who demonstrated, in somatic cell hybridization studies between SV40-immortalized cells, complementation to senescence despite continued SV40 T antigen expression.[94] Similar data were reported by Chen and colleagues for HPV-immortalized cells.[95] These results suggested the existence of an intracellular control mechanism acting at the functional level of viral oncoproteins.[96]

An additional and apparently independent signaling pathway regulating the transcription of "high risk" HPVs was demonstrated by Rösl et al[97-99] and by further studies conducted by Bartsch et al[100] and Dürst and colleagues.[101] In essence these data revealed that in immortalized cells upon inoculation into a suitable host viral transcriptional activity is rapidly shut-off and, as Rösl et al could demonstrate, this is mediated by paracrine regulation, apparently via TNF and other cytokines released from macrophages and possibly from other cells.[99] In malignant cells this regulatory pathway appears to be generally interrupted; they do not reveal the same transcriptional repression. Data from Smits et al suggests an involvement of a locus on the short arm of chromosome 11 and of protein phosphatase 2A in the transcriptional control of "high risk" HPV expression.[102]

The induction of an interesting cytokine, the macrophage-chemoattractant protein-1 (MCP-1), selectively by immortalized but not by malignant cells tested thus far, may point to a new mechanism of immunosuppression by "high risk" HPV oncogenes.[99] Whereas herpes and adenovirus infections interfere with epitope presentation and MHC-transport proteins, a suppression of MCP-1 may specifically perturb a macrophage-mediated control.[103,104]

Thus, progression toward malignancy of "high risk" HPV-infected cells appears to require the interruption of two independent signaling pathways, the interruption of dichotomous control, one regulating the function of viral oncoproteins, the other suppressing viral transcription.[96]

Substantial excitement was created by the discovery that the E7 protein of "high risk" viruses binds the cellular pRB protein, disrupting pRB interaction with the cellular transcription factor E2F and the E6 binds the p53 protein, known to be modified in many human cancers.[105-107] This was based on earlier observations reporting similar properties of SV40 and adenovirus oncoproteins.[108-110] The binding of p53 added an important aspect to our present understanding of the mechanism by which these viruses contribute to oncogenesis. The rapid degradation of p53 by E6 as shown by Scheffner et al[111] seems to represent one interaction contributing to genomic instability, however, it is likely not the only one since E7 also leads to chromosomal instability.[112]

These observations offered the possibility to functionally explain the term "high risk" HPV. The induction of chromosomal instability seems to be the precondition for mutations in specific signaling pathways and provides a reasonable explanation for "high risk" HPVs acting as solitary carcinogens.[113] Low risk HPVs either do not bind p53 or do it at lower affinity. Based on the presence of those HPV types in cancer biopsies, the original clinically oriented definition of "high and low risk" HPVs now finds functional support.[83] Additional data further underlining this concept were obtained by Hurlin et al and Pecoraro and colleagues who reported the spontaneous malignant conversion of HPV 18-immortalized human keratinocyte lines after long-time cultivation in vitro without exposing them to other carcinogens.[114,115]

Low risk viruses seem to require external cofactors (mutagens) for the rare induction of malignant tumors.[113]

The interest of present and future studies will likely focus on viral oncoprotein interactions with host cell proteins regulating the cell cycle. Interesting data are forthcoming, particularly in interactions with cyclins A and E.[116,117] The expression of viral oncoproteins in dependence on cell types and differentiation will be studied in more detail in transgenic animals. Other future research will analyze cellular genes within signaling cascades controlling viral oncoprotein function or expression. This may have major impact on our understanding of the mechanisms in viral carcinogenesis. A further promising aspect, basically resulting from these studies, will be the application of this knowledge to early diagnosis, therapy and particularly to prevention of HPV-linked diseases. It still seems a long way to successful gene therapeutic approaches. The development of serological methods, however, for the detection of previous HPV exposures based on the synthesis of virus-like particles in recombinant systems,[118,119] and particularly their use as vaccines for prevention of HPV infections, look remarkably promising.

Although epidemiological and seroepidemiological studies initially linked hepatitis viruses with liver cancer, Epstein-Barr virus with B cell lymphomas and *Helicobacter pylori* infections with gastric cancer (see IARC monographs, refs. 120, 121), they entered the HPV field relatively late. It took a number of years to validate the HPV test systems and to assess the credibility of certain laboratory results. In recent years this situation has changed, however, and "high risk" HPV infections have been established as the sole or at least predominant risk factor for cervical cancer development.[122-124] The findings of specific HPV DNA in more than 96% of cervical premalignant lesions when testing permitted the identification of more than 60 of the known HPV types[125] and, with slightly less sensitive technology, in 92% of cervical cancer biopsies[124] is driving the detection systems to the borderlines of their sensitivity.

A number of questions must still be answered before we fully understand the mechanism of HPV-induction of cancer of the cer-

vix. Yet, more than 150 years after Rigoni-Stern[67] and 13 years after identifying "high risk" viruses in cervical cancer, the quest for a causal role of a specific infectious agent appears to be clarified today, at least for the vast majority of cancers of the cervix. It is safe to state that the second most frequent cancer in women worldwide and the most frequent gynecological disorder, cervical intraepithelial neoplasia, have a papillomavirus etiology.

1.4.4. Other nonanogenital cancers

The role of HPV in other human cancers, besides those of the skin and the anogenital tract, is an exciting topic today. Cancers of the esophagus, as initially postulated by Syrjänen,[126] cancers of the oropharyngeal and respiratory routes, even neoplasms of internal organs and systemic malignancies deserve to be tested in this respect. The first promising results from a low percentage of cancers of the tongue and the oral cavity,[127,128] of cancers of the lung[129] and of laryngeal cancers[130,131] date back more than 10 years. There still exists the need for a systematic and comprehensive approach, taking into account all existing HPV types and the possibility of yet unknown, distantly related novel types. Within the last year we experienced an hitherto unanticipated blossoming in the identification of putative new types. Within this period Shamanin and de Villiers and colleagues identified 25 putative novel HPV types in our laboratory (de Villiers et al, unpublished results and refs. 35, 132), most of them not yet fully characterized. If we include those in the number of presently known HPV genotypes, we reach a count exceeding 100 genotypes of human pathogenic papillomaviruses. It is likely this number will increase as we look into extragenital and noncutaneous sites.

Unquestionably, the field is complex and complicated. The emerging picture of a relationship of about 15% of the worldwide cancer burden (by including nonmelanoma skin cancers) solely to papillomavirus infections, with the prospect of increasing this percentage further in the future, renders it most fascinating and classifies it as an area of important clinical impact and high medical priority.

REFERENCES

1. Iwanowski D. Über die Mosaikkrankheit der Tabakpflanze. Bull Acad Imp Sci St Petersburg 1894; 3(35):67-70.
2. Beijerinck MW. Über ein Contagium vivum fluidum als Ursache der Fleckenkrankheit der Tabakblätter. Verhandl. Koninkl Akad Wetenschappen te Amsterdam 1898; 6:3-22.
3. McFadyan J, Hobday F. Note on the experimental "transmission of warts in the dog". J Comp Pathol Ther 1898; 11:341-349.
4. Ciuffo G. Innesto positivo con filtrado di verrucae volgare. G Ital Mal Venereol 1907; 48:12-18.
5. Rowson KEK, Mahy BWJ. Human papova (wart) virus. Bacteriol Rev 1967; 31:110-131.
6. Shope RE. Infectious papillomatosis of rabbits. J Exp Med 1933; 58:607-627.
7. Rous P, Beard JW. Carcinomatous changes in virus-induced papillomas of the skin of the rabbit. Proc Soc Exp Biol Med 1934; 32:578-580.
8. Rous P, Beard JW. The progression to carcinoma of virus-induced rabbit papilloma (Shope). J Exp Med 1935; 62:523-548.
9. Syverton JT, Berry GP. Carcinoma in the cottontail rabbit following spontaneous virus papilloma (Shope). Proc Soc Exp Biol Med 1935; 33:300-400.
10. Rous P, Kidd JG. The carcinogenic effect of a papillomavirus on the tarred skin of rabbits. I. Description of the phenomenon. J Exp Med 1938; 67:399- 422.
11. Rous P, Friedewald WF. The effect of chemical carcinogens on virus-induced rabbit papillomas. J Exp Med 1944; 79:511-537.
12. Strauss MJ, Shaw EW, Bunting H et al. "Crystalline" virus-like particles from skin papillomas characterized by intranuclear inclusion bodies. Proc Soc Exp Biol Med 1949; 72:46-50.
13. Crawford LV, Crawford EM. A comparative study of polyoma and papilloma viruses. Virology 1963; 21:258-263.
14. Ito Y, Evans CA. Induction of tumors in domestic rabbits with nucleic acid preparations from partially purified Shope papilloma virus and from extracts of the papillomas of domestic and cotton tail rabbits. J Exp Med 1961; 114:485-491.
15. Han RF, Breitburd F, Marche PN et al. Linkage of regression and malignant conversion of rabbit viral papillomas to MHC class II genes. Nature 1992; 356:66-68.
16. Zeltner R, Borenstein LA, Wettstein FO et al. Changes in RNA expression pattern during malignant progression of cottontail rabbit papillomavirus-induced tumors in rabbits. J Virol 1994; 68:3620-2630.
17. Lin YL, Borenstein LA, Ahmed R et al. Cottontail rabbit papillomavirus L1 protein-based vaccines: protection is achieved only with a full-length, nondenatured product. J Virol 1993; 67:4154-4163.

18. Breitburd F, Kirnbauer R, Hubbert NL et al. Immunization with virus-like particles from cottontail rabbit papillomavirus (CRPV) can protect against experimental CRPV infection. J Virol 1995; 69: 3959-3963.

19. Magelhaes O. Verruga dos bovideos. Brasil-Medico 1920; 34:430-431.

20. Olson C, Cook RH. Cutaneous sarcoma-like lesions of the horse induced by the agent of bovine papilloma. Proc Soc Exp Biol Med 1951; 77:281-284.

21. Olson C, Pamukcu AM, Brobst DF et al. A urinary bladder tumor induced by a bovine cutaneous papilloma agent. Cancer Res 1959; 19:779-782.

22. Friedmann JC, Levy JP, Lasnaret J et al. Induction de fibromes sous-cutanés chez le hamster doré par inoculation d'extrait acellulaires de papillomes bovins et leur transformation maligne par greffes isologues. Compt Rend Acad Sci (Paris) 1963; 257:2328-2331.

23. Boiron M, Levy JP, Thomas M et al. Some properties of bovine papilloma virus. Nature 1964; 201:423-424.

24. Black PH, Hartley JW, Rowe WP et al. Transformation of bovine tissue culture cells by bovine papilloma virus. Nature 1963; 199:1016-1018.

25. Thomas M, Levy JP, Tanzer J et al. Transformation in vitro de cellules de peau de veau embryonnaire sous l'action d'extraits acellulaires de papillomes bovins. Compt Rend Acad Sci (Paris) 1963; 257: 2155-2158.

26. Lancaster WD, Olson C. Demonstration of two distinct classes of bovine papilloma virus. Virology 1978; 89:371-379.

27. Lowy DR, Dvoretzky I, Shober R et al. In vitro tumorigenic transformation by a defined subgenomic fragment of bovine papilloma virus DNA. Nature 1980; 287:72-74.

28. Chen EY, Howley PM, Levinson AD et al. The primary structure and genetic organization of the bovine papillomavirus (BPV) type 1 genome. Nature 1982; 299:529-534.

29. Jarrett WFH. Bracken fern and papilloma virus in bovine alimentary cancer. Brit Med Bull 1980; 36:79-81.

30. Almeida JD, Goffe AP. Antibody to wart virus in human sera demonstrated by electron microscopy and precipitin tests. Lancet 1965; 2:1205-1207.

31. zur Hausen H, Meinhof W, Scheiber W et al. Attempts to detect virus-specific DNA sequences in human tumors: I. Nucleic acid hybridizations with complementary RNA of human wart virus. Int J Cancer 1974; 13:650-656.

32. zur Hausen H, Schulte-Holthausen H, Wolf H et al. Attempts to detect virus-specific DNA in human tumors: II. Nucleic acid hybridizations with complementary RNA of human herpes group viruses. Int J Cancer 1974; 13:657-664.

33. Gissmann L, zur Hausen H. Human papilloma viruses: physical mapping and genetic heterogeneity. Proc Nat Acad Sci USA 1976; 73:1310-1313.
34. Gissmann L, Pfister H, zur Hausen H. Human papilloma viruses (HPV): Characterization of four different isolates. Virology 1977; 76:569-580.
35. Orth G, Favre M, Croissant O. Characterization of a new type of human papillomavirus that causes skin warts. J Virol 1977; 24: 108-120.
36. Coggin JR, zur Hausen H. Workshop on papilloma viruses and cancer. Cancer Res 1978; 39:545-546.
37. de Villiers EM. Human pathogenic papillomaviruses: An update. In: zur Hausen H, ed. Current Topics in Microbiology and Immunology. Berlin-Heidelberg: Springer Verlag, 1994; 86:1-12.
38. Shamanin V, Glover M, Rausch C et al. Specific types of human papillomavirus found in benign proliferations and carcinomas of the skin in immunosuppressed patients. Cancer Res 1994; 54:4610-4613.
39. Van Ranst M, Fuse A, Fiten P et al. Human papillomavirus type 13 and pygmy chimpanzee papillomavirus type 1: comparison of the genome organizations. Virology 1992; 190:587-596.
40. Bernard HU, Chan SY, Delius H. Evolution of papillomaviruses. Curr Topics Microbiol Immunol 1994; 186:33-54.
41. Lewandowsky F, Lutz W. Ein Fall einer bisher nicht beschriebenen Hauterkrankung (Epidermodysplasia verruciformis). Arch Dermatol Syph (Berlin) 1922; 141:193-203.
42. Schellender F, Fritsch F. Epidermodysplasia verruciformis. Neue Aspekte zur Symptomatologie und Pathogenese. Dermatologica 1970; 140:251-259.
43. Ruiter M and van Mullem PJ. Behaviour of virus in malignant degeneration of skin lesions in epidermodysplasia verruciformis. J Invest Dermatol 1970; 54:324-331.
44. Jablonska S, Dabrowski J, Jakubowicz K. Epidermodysplasia verruciformis as a model in studies on the role of papovaviruses in oncogenesis. Cancer Res 1972; 32:583-589.
45. Lutzner MA. Epidermodysplasia verruciformis. An autosomal recessive disease characterized by viral warts and skin cancer. Bull Cancer Paris 1978; 65:169-182.
46. Lutz W. A propos de l'epidermodysplasie verruciforme. Dermatologica 1946; 92:30-43.
47. Jablonska S, Millewski B. Zur Kenntnis der Epidermodysplasia verruciformis Lewandowsky-Lutz. Dermatologica 1957; 115:1-22.
48. Ruiter M, van Mullem PJ. Demonstration by electronmicroscopy of an intranuclear virus in epidermodysplasia verruciformis. J Invest Dermatol 1966 Sep; 47(3):247-52.

49. Yabe Y, Okamoto T, Okmori S et al. Virus particles in epidermodysplasia verruciformis with carcinoma. Dermmatologica 1969; 139:161-164.
50. Delescluse C, Prunieras M, Regnier M et al. Epidermodysplasia verruciformis. I. Electron microscope autoradiography and tissue culture studies. Arch Dermatol Forsch 1972; 242:202-215.
51. Orth G, Jablonska S, Jarzabek-Chorzelska M et al. Characteristics of the lesions and risk of malignant conversion as related to the type of the human papillomavirus involved in epidermodysplasia verruciformis. Cancer Res 1979; 39:1074-1082.
52. Preston DS, Stern RS. Nonmelanoma cancers of the skin. New Engl J Med 1992; 327:1649-1662.
53. Rees J. Genetic alterations in nonmelanoma skin cancer. J. Invest Dermatol 1994; 103:747-750.
54. Lutzner MA, Orth G, Dutronquay V et al. Detection of human papillomavirus type 5 DNA in skin cancers of an immunosuppressed renal allograft recipient. Lancet 1983; 2:422-424.
55. Obalek S, Favre M, Szymanczyk J et al. Human papillomavirus (HPV) types specific of epidermodysplasia verruciformis in warts induced by HPV 3 or HPV3-related types in immunosuppressed patients. J Invest Dermatol 1992; 98:936-941.
56. Van der Leest RJ, Zachow KR, Ostrow RS et al. Human papillomavirus heterogeneity in 36 renal transplant recipients. Arch Dermatol 1986; 123:354-357.
57. Euvrard S, Chardonnet Y, Pouteil-Noble C et al. Association of skin malignancies with various and multiple carcinogenic and noncarcinogenic human papillomaviruses in renal transplant recipients. Cancer 1993; 72:2198-2206.
58. Stark LA, Arends MJ, McLaren KM et al. Prevalence of human papillomavirus DNA in cutaneous neoplasms from renal allograft recipients supports a possible viral role in tumor promotion. Brit J Cancer 1994; 69:222-229.
59. Barr BB, Benton EC, McLaren K et al. Human papilloma virus infection and skin cancer in renal allograft recipients. Lancet 1989; 1:124-129.
60. Purdie KJ, Sexton CJ, Proby CM et al. Malignant transformation of cutaneuos lesions in renal allograft patients: a role for human papillomavirus. Cancer Res 1993; 53:5328-5333.
61. Soler C, Chardonnet Y, Allibert P et al. Detection of mucosal human papillomavirus types 6/11 in cutaneous lesions from transplant recipients. J Invest Dermatol 1993; 101:286-291.
62. Berkhout RJM, Tieben LM, Smits HL et al. Detection and typing of epidermodysplasia verruciformis-associated human papillomavirus types in cutaneous cancers from renal transplant recipients: a nested approach. J Clin Microbiol 1965; 33:690-695.

63. Grimmel M, de Villiers EM, Pawlita M et al. Characterization of a new human papillomavirus type (HPV 41) isolated from dissiminated warts and the detection of closely related sequences in some squamous cell carcinomas. Int J Cancer 1988; 41:5-9.

64. Kawashima M, Favre M, Obalek S et al. Premalignant lesions and cancers of the skin in the general population: evaluation of the role of human papillomaviruses. J Invest Dermatol 1990; 95:537-42.

65. Moy RL, Eliezri YD, Nuovo GJ. HPV DNA in periungual SSC. J Am Med Assoc 1989; 261:2669-2673.

66. Eliezri YD, Silverstein SJ, Nuovo GJ. Occurrence of human papillomavirus type 16 DNA in cutaneous squamous and basal cell neoplasms. J Am Acad Dermatol 1990; 23:836-842.

67. Rigoni-Stern D. Fatti statistici relativialle malatia cancerose. G Serv Prog Pathol Therap 1842; 2:507-517.

68. Rotkin ID. A comparison review of key epidemiological studies in cervical cancer related to current searches for transmissible agents. Cancer Res 1973; 33:1353-1367.

69. zur Hausen H. Condylomata acuminata and human genital cancer. Cancer Res 1976; 36:530.

70. zur Hausen H. Human papillomaviruses and their possible role in squamous cell carcinomas. Current Topics Microbiol Immunol 1977; 78:1-30.

71. Meisels A, Fortin R. Condylomatous lesions of the cervix and vagina. I. Cytological patterns. Act Cytologica 1976; 20:505-509.

72. Meisels A, Roy M, Fortier M et al. Human papillomavirus infection of the cervix: the atypical condyloma. Act Cytologica 1981; 25:7-16.

73. Gissmann L, zur Hausen H. Partial characterization of viral DNA from human genital warts (condylomata acuminata). Int J Cancer 1980; 25:605-609.

74. de Villiers EM, Gissmann L, zur Hausen H. Molecular cloning of viral DNA from human genital warts. J Virol 1981; 40:932-935.

75. Gissmann L, Diehl V, Schultz-Coulon H et al. Molecular cloning and characterization of human papillomavirus DNA from a laryngeal papilloma. J Virol 1982; 44:393-400.

76. Dürst, M., Gissmann, L., Ikenberg, H., and zur Hausen, H. 1983. A papillomavirus DNA from a cervical carcinoma and its prevalence in cancer biopsy samples from different geographic regions. Proc Nat Acad Sci USA 80, 3812-3815.

77. Boshart M, Gissmann L, Ikenberg H et al. A new type of papillomavirus DNA, its presence in genital cancer and in cell liness derived from genital cancer. EMBO J 1984; 3:1151-1157.

78. Ikenberg H, Gissmann L, Gross G et al. Human papillomavirus type 16 related DNA in genital Bowen's disease and in Bowenoid papulosis. Int J Cancer 1983; 32:563-564.

79. Schwarz E, Freese K, Gissmann L et al. Structure and transcription of human papillomvirus sequences in cervical carcinoma cells. Nature 1985; 314:111-114.
80. zur Hausen H. Cell-virus gene balance hypothesis of carcinogenesis. Behring Inst Mitt 1977; 61:23-30.
81. zur Hausen H. The role of viruses in human tumors. In: Klein G, Weinhouse S, eds. Advances in Cancer Res 1980; 33:77-107.
82. zur Hausen H. Intracellular surveillance of persisting viral infections: Human genital cancer resulting from failing cellular control of papillomavirus gene expression. Lancet 1986; 2:489-491.
83. zur Hausen H. Genital papillomavirus infections. In: Rigby PWJ, Wilkie NM, eds. Viruses and Cancer. Cambridge: Cambridge University Press 1986; 83-90.
84. zur Hausen H. Papillomaviruses in anogenital cancer: A model to understand the role of viruses in human cancers. Cancer Res 1989; 49:4677-4681.
85. Yasumoto S, Burckhardt AL, Doninger J et al. Human papillomavirus type 16 DNA induced malignant transformation of NIH3T3 cells. J Virol 1986; 57:572-577.
86. Dürst M, Dzarlieva-Petrusevska RT, Boukamp P et al. Molecular and cytogenetic analysis of immortalized human primary keratinocytes obtained after transfection with human papillomavirus type 16 DNA. Oncogene 1987; 1:251-256.
87. Pirisi L, Yasumoto S, Fellery M et al. Transformation of human fibroblasts and keratinocytes with human papillomavirus type 16 DNA. J Virol 1987; 61:1061-1066.
88. Münger K, Phelps WC, Bubb V et al. The E6 and E7 genes of human papillomavirus type 16 are necessary and sufficient for transformation of primary human keratinocytes. J Virol 1989; 63:4417-4423.
89. Halbert CL, Demers GW, Galloway DA. The E7 gene of human papillomavirus type 16 is sufficient for immortalization of human epithelial cells. J Virol 1991; 65:473-478.
90. Band V, De Caprio JA, Delmolina L et al. Loss of p53 protein in human papillomavirus type 16 E6-immortalized human mammary epithelial cells. J Virol 1991; 65:6671-6676.
91. von Knebel Doeberitz M, Oltersdorf T, Schwarz E et al. Correlation of modified human papillomavirus early gene expression with altered growth properties in C4-1 cervical carcinoma cells. Cancer Res 1988; 48:3780-3786.
92. von Knebel Doeberitz M, Rittmüller C, zur Hausen H et al. Inhibition of tumorigenicity of cervical cancer cells in nude mice by HPV E6-E7 antisense RNA. Int J Cancer 1992; 51:831-834.
93. von Knebel Doeberitz M, Rittmüller C, Aengeneyndt F et al. Reversible repression of papillomavirus oncogene expression in cervical carcinoma cells: consequences for the phenotype and E6-p53 and E7-pRB interactions. J Virol 1994; 68:2811-2821.

94. Pereira-Smith OM, Smith JR. Expression of SV40 antigen in finite lifespan hybrids of normal and SV40-transformed fibroblasts. Somatic Cell Genetics 1981; 7:411-421.

95. Chen TM, Pecoraro G, Defendi V. Genetic analysis of in vitro progression of human papillomavirus-transfected human cervical cells. Cancer Res 1993; 53:1167-1171.

96. zur Hausen H. Disrupted dichotomous intracellular control of human papillomavirus infection in cancer of the cervix. Lancet 1994; 343:955-957.

97. Rösl F, Dürst M, zur Hausen H. Selective suppression of human papillomavirus transcription in nontumorigenic cells by 5-azacytidine. EMBO J 1988; 7:1321-1328.

98. Rösl F, Achtstetter T, Hutter KJ et al. Extinction of the HPV 18 upstream regulatory region in cervical carcinoma cells after fusion with nontumorigenic human keratinocytes under nonselective conditions. EMBO J 1991; 10:1337-1345.

99. Rösl F, Lengert M, Albrecht J et al. Differential regulation of the JE gene encoding the monocyte chemoattractant protein (MCP-1) in cervical carcinoma cells and derived hybrids. J Virol 1994; 68: 2142-2150.

100. Bartsch D, Boye B, Baust C et al. Retinoic acid-mediated repression of human papillomavirus 18 transcription and different ligand regulation of the retinoic acid receptor β gene in nontumorigenic and tumorigenic HeLa hybrid cells. EMBO J 1992; 11:2283-2291.

101. Dürst M, Glitz D, Schneider A et al. Human papillomavirus type 16 (HPV 16) gene expression and DNA replication in cervical neoplasia: analysis by in situ hybridization. Virology 1992; 189:132-140.

102. Smits PHM, Smits HL, Minnaar R et al. The trans-activation of the HPV 16 long control region in human cells with a deletion in the short arm of chromosome 11 is mediated by the 55kDa regulatory subunit of protein phosphatase 2A. EMBO J 1992; 11:4601-4606.

103. York IA, Roop C, Andrews DW et al. A cytosolic Herpes simplex virus protein inhibits antigen presentation to CD8+ T lymphocytes. Cell 1994; 77:525-535.

104. Ge R, Liu X, Ricciardi RP. E1A oncogene of adenovirus-12 mediates trans-repression of MHC class I transcription in Ad5/Ad12 somatic hybrid transformed cells. Virology 1994; 203:389-392.

105. Dyson N, Howley PM, Münger K et al. The human papillomavirus-16 E7 oncoprotein is able to bind to the retinoblastoma gene product. Science 1989; 243:934-937.

106. Chellappan S, Kraus V, Kroger B et al. Adenovirus E1A, simian virus 40 tumor antigen, and human papillomavirus E7 protein share the capacity to disrupt the interaction between transcription factor E2F and the retinoblastoma gene product. Proc Nat Acad Sci USA 1992; 89:4549-4553.

107. Werness BA, Levine AJ, Howley PM. Association of human papillomavirus types 16 and 18 E6 proteins with p53. Science 1990; 248:76-79.
108. Lane DP, Crawford LV. T antigen is bound to a host protein in SV 40-transformed cells. Nature 1979; 278:261-263.
109. Linzer DIH, Levine AJ. Characterization of a 54K dalton cellular SV40 tumor antigen present in SV40-transformed cells and uninfected embryonal carcinoma cells. Cell 1979; 17:43-52.
110. DeCaprio JA, Ludlow JW, Figge J et al. SV40 large tumor antigen forms a specific complex with the product of the retinoblastoma susceptibility gene. Cell 1988; 54:275-283.
111. Scheffner M, Werness BA, Huibregtse JM et al. The E6 oncoprotein encoded by human papillomavirus types 16 and 18 promotes the degradation of p53. Cell 1990; 63:1129-1136.
112. White AE, Livanos EM, Tlsty TD. Differential disruption of genomic integrity and cell cycle regulation in normal human fibroblasts by the HPV oncoproteins. Genes and Development 1994; 8:666-677.
113. zur Hausen H. Viruses in human cancer. Science 1991; 254:1167-1173.
114. Hurlin PJ, Kaur P, Smith P et al. Progression of human papillomavirus type 18 immortalized human keratinocytes to a malignant phenotype. Proc Natl Acad Sci USA 1981; 88:570-574.
115. Pecoraro G, Lee M, Morgan D et al. Evolution of in vitro transformation and tumorigenesis of HPV 16 and HPV 18 immortalized primary cervical epithelial cells. Am J Pathol 1991; 138:1-8.
116. Pagano M, Dürst M, Joswig S et al. Binding of the human E2F transcription factor to the retinoblastoma protein but not to cyclin A is abolished in HPV-16-immortalized cells. Oncogene 1992; 7:1681-1686.
117. Zerfass K, Schulze A, Spitkovsky D et al. S phase induction by the human papillomavirus 16 E7 oncogene reveals an activation cascade of cyclin E and cyclin A transcription. J Virol 1995; 69:6389-6399.
118. Kirnbauer R, Booy F, Cheng N et al. Papillomavirus L1 major capsid protein self-assembles into virus-like particles that are highly immunogenic. Proc Natl Acad Sci USA 1992; 89:12180-12184.
119. Kirnbauer R, Hubbert NL, Wheeler CM et al. A virus-like particle enzyme-linked immunosorbent assay detecs serum antibodies in a majority of women infected with human papillomavirus type 16. J Nat Cancer Inst 1994; 86:494-9.
120. IARC Monograph on Evaluation of Carcinogenic Risks of Humans. Vol. 59. Hepatitis Viruses, IARC Lyon, 1994.
121. IARC Monograph on Evaluation of Carcinogenic Risks of Humans. Vol. 61. Schistosomes, Liver Flukes and *Helicobacter pylori*. IARC Lyon, 1994.
122. Muñoz N, Bosch FX, de Sanjose S et al. The causal link between human papillomavirus and invasive cervical cancer: a population-based case-control study in columbia and Spain. Int J Cancer 1992; 52:743-749.

123. Schiffman MH, Bauer HM, Hoover RN et al. Epidemiological evidence showing that human papillomavirus infection causes most cervical intraepithelial neoplasia. J Natl Cancer Inst 1993; 85:958-964.
124. Bosch FX, Manos MM, Muñoz N et al. Int Biol Study Cervical Cancer (IBSSC) Study Group. Prevalence of human papillomavirus in cervical cancer: a worldwide perspective. J Natl Cancer Inst 1995; 87:796-802.
125. Matsukura T, Sugase M. Identification of genital human papillomaviruses in cervical biopsy specimen: segregation of specific virus types in specific clinicopathologic lesions. Int J Cancer 1995; 61:13-22.
126. Syrjänen K. Histological changes identical to those of condylomatous lesions found in esophageal squamous cell carcinomas. Arch Geschwulstforsch 1982; 52:283-292.
127. de Villiers EM, Weidauer H, Otto H et al. Papillomavirus DNA in human tongue carcinomas. Int J Cancer 1985; 36:575-578.
128. Löning T, Ikenberg H, Becker J et al. Analysis of oral papillomas, leukoplakias and invasive carcinomas for human papiloomavirus type related DNA. J Invest Dermatol 1985; 84:417-420.
129. Stremlau A, Gissmann L, Ikenberg H et al. Human papilloma virus type 16 related DNA in an anaplastic carcinoma of the lung. Cancer 1995; 55:1737-1740.
130. Kahn T, Schwarz E, zur Hausen H. Molecular cloning and characterization of the DNA of a new human papillomavirus (HPV 30) from a laryngeal carcinoma. Int J Cancer 1986; 37:61-65.
131. Scheurlen W, Stremlau A, Gissmann L. Rearranged HPV 16 molecules in an anal and in a laryngeal carcinoma. Int J Cancer 1986; 38:671-676.
132. Shamanin V, zur Hausen H, Lavergne D et al. HPV infections in nonmelanoma skin cancers from renal transplant recipients and nonimmunosuppressed patients. J Nat Cancer Inst 1996; 88:802-811.
133. De Villiers E M. 1989. Heterogeneity of human papillomavirus group. J Virol 63:4898-4903.
134. De Villiers EM. Papillomavirus and HPV typing. Clinics in Dermatology 1997; (in press).

CHAPTER 2

Regulation of E6 and E7 Oncogene Transcription

Frank Rösl and Elisabeth Schwarz

2.1. INTRODUCTION

Expression of the viral oncogenes E6 and E7 is fundamental to HPV-associated carcinogenesis. The multistep progression of a persistent infection by HPV16 or HPV18 or another "high risk" HPV type from a clinically inconspicuous state to a detectable precursor lesion and eventually to a carcinoma is thought to be driven mainly by the oncoproteins E6 and E7. Furthermore, the maintenance of the malignant phenotype also requires the continuous expression of the HPV oncogenes.

The human papillomaviruses produce persistent infections of squamous and mucous epithelia and induce hyperproliferation of the infected cells. For the successful establishment of a persistent infection, the viruses must infect the basal cells because these are the only cells in the epithelium which are capable of dividing. The basal cells produce all the cells of the upper layers, which in the course of differentiation lose their capacity for cell division and eventually are shed from the epithelium. Vegetative viral DNA replication, expression of the late genes and virion assembly are closely linked to the differentiation state of the host cells and occur only

Papillomaviruses in Human Cancer: The Role of E6 and E7 Oncoproteins, edited by Massimo Tommasino. © 1997 Landes Bioscience.

in the differentiated keratinocytes of the uppermost cell layers. It is a major challenge of HPV research to identify the factors which cause the differentiation dependence of HPV gene expression and replication.

In the initial phase of infection, after entry of the virion DNA into the cell nucleus, transcription of the HPV early genes is exclusively dependent on cellular transcription factors. In the later phases, virus-specific proteins participate in the regulation of both early and late viral gene expression. Work over the last decade has led to the identification of many of the cellular factors which regulate transcription of the E6 and E7 oncogenes. At least one viral factor, namely E2, also plays a crucial role.

Deregulation of E6 and E7 gene expression, in particular the escape from cellular negative control mechanisms, is thought to be important for the carcinogenic progression of HPV-infected cells. Several putative repressors which have been identified seem to be involved in this process. It is a task of basic research to unravel the factors involved in the regulation of E6 and E7 oncogene expression and their alterations associated with carcinogenesis. In addition, this gained knowledge has potential clinical impact, because it may provide a basis for the development of therapeutic strategies to inhibit the expression of the E6 and E7 oncogenes and thus to block tumor growth.

The first part of this chapter summarizes the data on the in vivo expression of the E6 and E7 oncogenes as determined by RNA-RNA in situ hybridization in precancerous lesions and carcinomas. In the second part, we will discuss viral and cellular transcription factors which are involved in the differential regulation of transcription of the E6 and E7 oncogenes. Since the viral DNA can persist in two physical states within the infected cells, either as extrachromosomal elements in a defined copy number or integrated into the host genome, the third part will discuss the consequences of viral DNA integration for E6 and E7 oncogene expression and the significance of such position effects during HPV-associated pathogenesis. Finally, we will describe experimental approaches and model systems which have been developed in order to analyze the differences of viral gene regulation in nontumorigenic and tumorigenic HPV-positive cells. This review will concentrate on the "high risk" prototypes HPV16 and HPV18 for which E6 and E7 oncogene expression has been studied most intensively. We refer only occasionally to other genital HPV types such as the "low risk" prototypes HPV6 and HPV11.

2.2. IN VIVO EXPRESSION OF THE E6 AND E7 ONCOGENES

Expression of the early and late papillomavirus genes in infected precancerous and malignant lesions has been intensively studied mainly by RNA-RNA in situ hybridization using subgenic probes of HPV16 or HPV18. In situ hybridizations on tissue sections of cervical intraepithelial neoplasias (CINs) and cervical carcinomas have given impressive insights into the topography of viral transcripts and DNA within the lesions, demonstrating that vegetative HPV DNA replication and the program of HPV gene expression are strictly dependent on the differentiation state of the host cells. Late gene transcripts are detectable only in the highly differentiated cells of the upper epithelial layers.[1-3] In contrast, transcripts for E6 and E7 are present in all cell layers, including the basal cell layer. In low grade CINs, the levels of E6/E7 transcripts are linked to the differentiation state of the host cells. The mRNA levels are very low in undifferentiated basal cells and are strongly increased in differentiated cells. Increased E6/E7 transcription is mainly found in those cells in which viral DNA replication and late gene expression take place, indicating that elevated E6/E7 RNA levels are due to the replication-dependent increase in the number of template DNA molecules.

High grade CIN lesions and cervical carcinomas tend to show high levels of E6/E7 transcripts which are evenly distributed throughout the undifferentiated epithelium.[2-4] Thus, in comparison to the low grade lesions, there seems to be an upregulation of E6 and E7 oncogene transcription in the high grade CINs and malignant lesions which may be due to the loss of negative cellular and viral controls. High grade CINs and invasive carcinomas show similar E6/E7 transcript patterns. This observation led to the important conclusion that further elevation of viral oncogene expression is not responsible for the progression to the malignant state. This indicates that additional alterations of cellular genes are required for full transformation to cervical cancer. Figures 2.1 and 2.2 show the in situ hybridization patterns of HPV16 E6/E7 transcripts in low and high grade CIN lesions (CIN I and CIN III) and in a cervical carcinoma.

The E6/E7 transcripts are usually spliced in the E6 gene region. Therefore they do not encode a full-length E6 protein, but rather contain a truncated E6* open reading frame (ORF).[5,6] As deduced from the 5'-ends of the E6/E7 transcripts, transcription is initiated at the early viral promoter located upstream of the E6 gene.

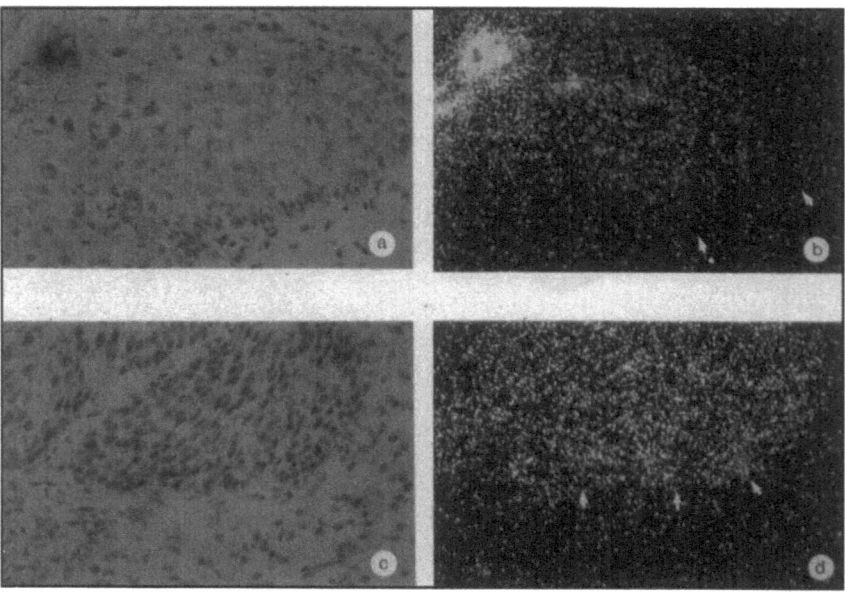

Fig. 2.1. HPV16 E6 and E7 oncogene transcripts in a low-grade and a high-grade cervical intraepithelial neoplasia (CIN I and CIN III) detected by RNA-RNA in situ hybridization. Brightfield (a and c) and darkfield (b and d) views are shown. The arrowheads in b and d mark the basement membrane. The pictures were kindly provided by M Dürst,[2] and reprinted with permission of Academic Press, Virology 1992; 189:132-140.

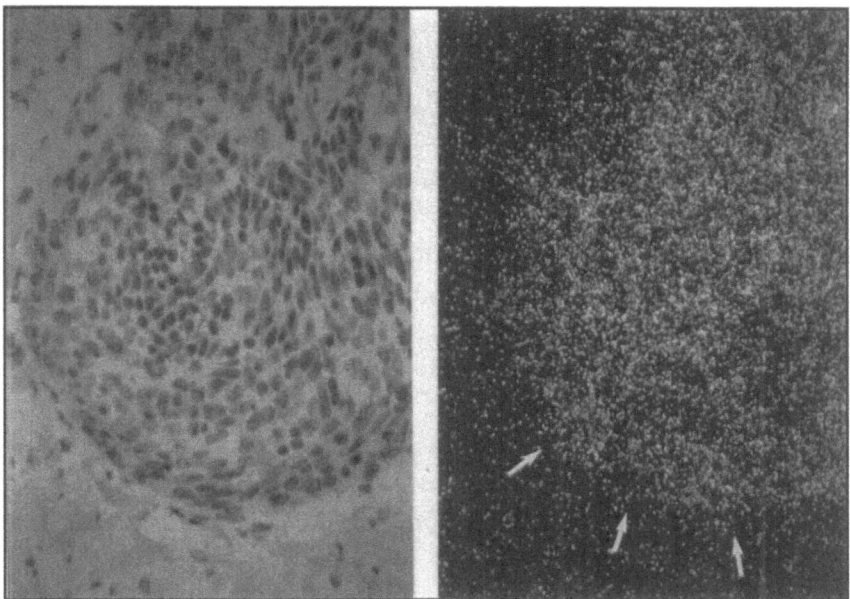

Fig. 2.2. HPV16 E6 and E7 oncogene transcripts in a cervical carcinoma detected by RNA-RNA in situ hybridization. Serial sections of the tumor are shown, stained with hematoxylin-eosin (left part) and after in situ hybridization (right part; darkfield view). The pictures were kindly provided by M Dürst,[2] and reprinted with permission of Academic Press, Virology 1992; 189:132-140.

2.3. VIRAL AND CELLULAR FACTORS FOR E6/E7 ONCOGENE TRANSCRIPTION

2.3.1. General aspects of transcription initiation

Initiation of transcription of protein-encoding genes is an intensively regulated step in eukaryotic gene expression. Signal transduction pathways converge at this point and allow the cell to adjust to the nutritional needs and the environmental information. Transcription is initiated after the basal transcription apparatus has assembled at the core promoter to build the pre-initiation complex. The basal apparatus is a multiprotein complex containing RNA polymerase II and the general transcription initiation factors TFIIA, -B, -D, -E, -F and -H.[7] The core promoters of many cellular and viral genes contain a TATA box element located about 25 to 30 bp upstream of the transcription start site. Binding of the TATA-binding protein (TBP) to the TATA box initiates the assembly process. TBP is part of the TFIID factor in which it is associated with several TBP-associated factors (TAFs). Until recently the TAFs were thought to be generally required for the activation of transcription. New data, however, indicate that this, at least in vivo, may not be the rule and that the TAFs are essential for cell cycle progression.[8] Furthermore, recent evidence indicates that in vivo RNA polymerase II is part of a large multiprotein complex termed RNA polymerase II holoenzyme in which it is assembled with the general transcription factors and additional components, including SRB proteins and the Swi/Snf complex capable of disrupting nucleosomes.[9,10] It is assumed that an initiation complex is formed by recruitment of the pre-assembled holoenzyme to the promoter.

Transcription initiation is activated by transcription factors that bind to enhancer elements that together build up the regulatory regions of the promoter.[11] The communication of the transcription factors with the basal transcription apparatus or the pol II holoenzyme occurs either directly or involves additional proteins termed coactivators. As targets of signal transduction pathways, the transcription factors are subject to modifications such as phosphorylation and dephosphorylation which determine their ability to interact with coactivators and thus their influence on the transcription rate. Regulation of transcription initiation also occurs by different mechanisms of repression which involve the inaccessibility of promoter regions to RNA polymerase in inactive chromatin, competition for DNA binding sites or steric hindrance between activators and repressors, heterodimerization of activators with inhibitory proteins, active repression by factors with intrinsic repressing activity and the inactivation of signal-targeted activators by dephosphorylation.[12]

2.3.2. Anatomy of the upstream regulatory region

The oncogenes E6 and E7 are located in the 5' part of the early region of the HPV genome and are followed by the genes E1,

E2 (which completely overlaps the small E4 gene) and E5. A poly(A)-signal sequence defines the 3'-end of the early region. The late region includes the two genes L2 and L1 which encode the capsid proteins. The part of the genome located between the genes L1 and E6 is called the upstream regulatory region (URR) because it contains essential regulatory elements for transcription of the early genes. Other names for the URR are "noncoding region" or NCR (because it is devoid of major open reading frames) and "long control region" or LCR. Since we will discuss this part of the HPV genome with regard to the transcriptional regulation of the downstream located E6 and E7 genes, we will use the term "upstream regulatory region" or URR. The URR has a size of approximately 800-1000 bp in the genomes of genital HPV types. It has been divided into three functionally distinct portions, the 5'-segment, the central segment and the 3'- segment. The composition of the URR of HPV16 and HPV18 is shown in Figure 2.3.

In the "high risk" genital HPV types, transcription of the oncogenes E6 and E7 is initiated at a promoter located in the 3'-segment of the URR just in front of the E6 gene. This promoter is called P_{97} in HPV16 and P_{105} in HPV18 (the subscript number indicates the nucleotide position of the most abundant 5'-end of the mRNA species initiated at the respective promoter). In this review, we will use the general term "E6 promoter" when no distinction between the HPV types is necessary. The primary transcripts initiated at the E6 promoter are terminated at the early poly(A)-signal. RNA processing gives rise to a complex set of alternatively spliced mRNAs with polycistronic structures.

The E6 core promoter, composed of a TATA box and a transcription start site, is under control of different enhancer elements which are located in the central and 3'-segment of the URR. They are recognized by sequence-specific transcription factors of either viral or cellular origin.

In addition to the E6 core promoter, the 3'-part of the URR contains two binding sites for the viral transcription factor E2, binding sites for the cellular transcription factors Sp1 and YY1, and the origin of replication with the binding site for the viral replication

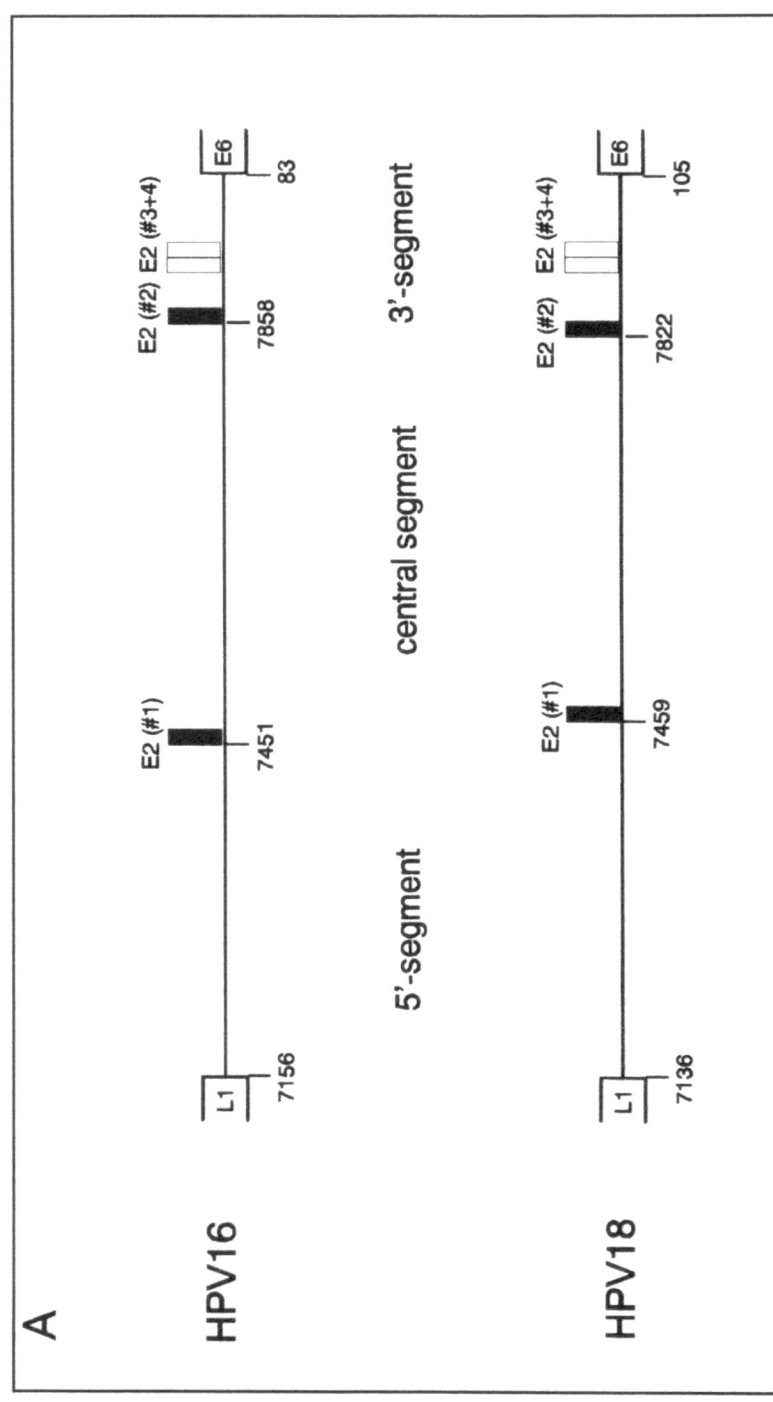

Fig. 2.3A (see legend, page 35).

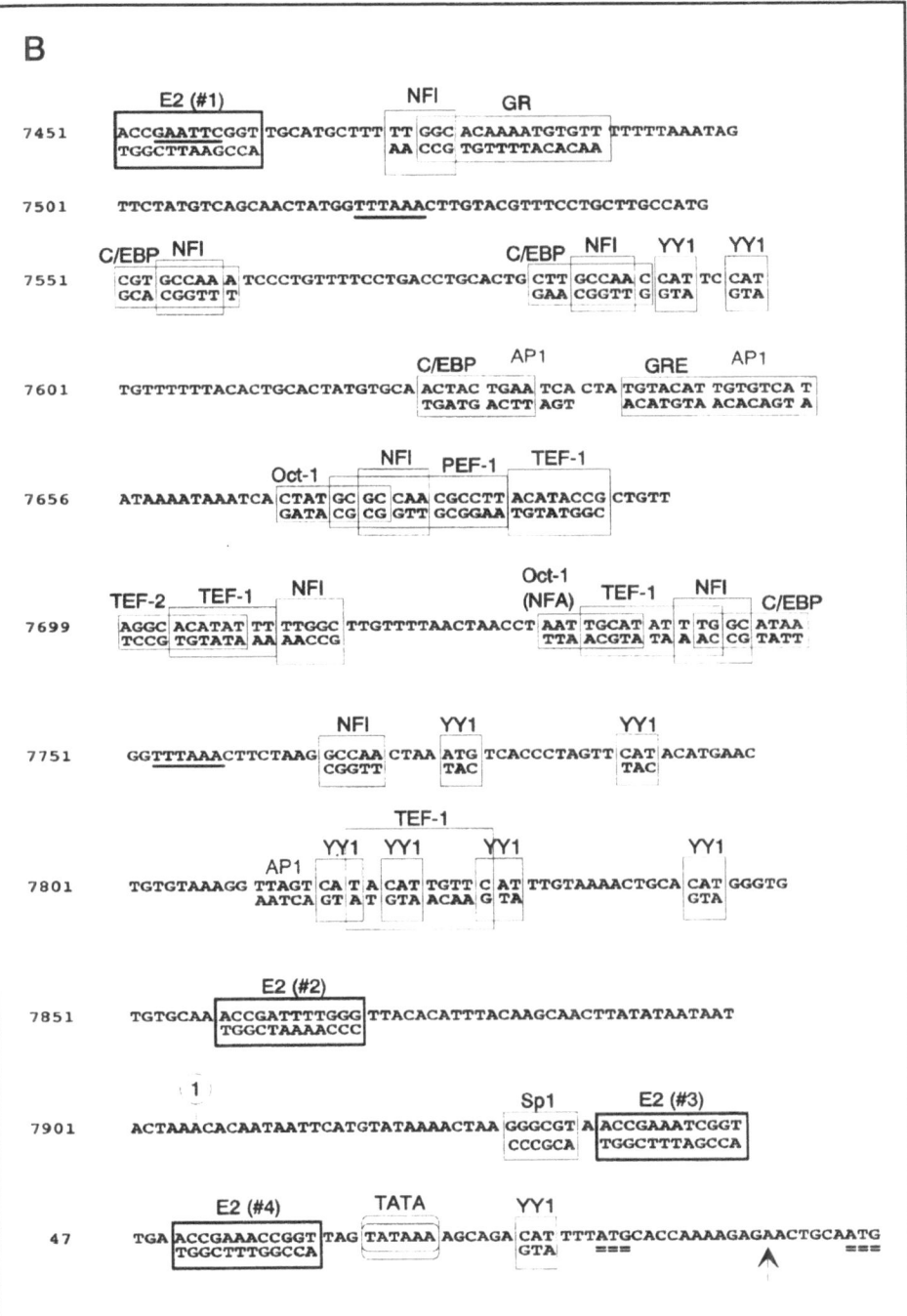

Fig. 2.3B (see legend, page 35).

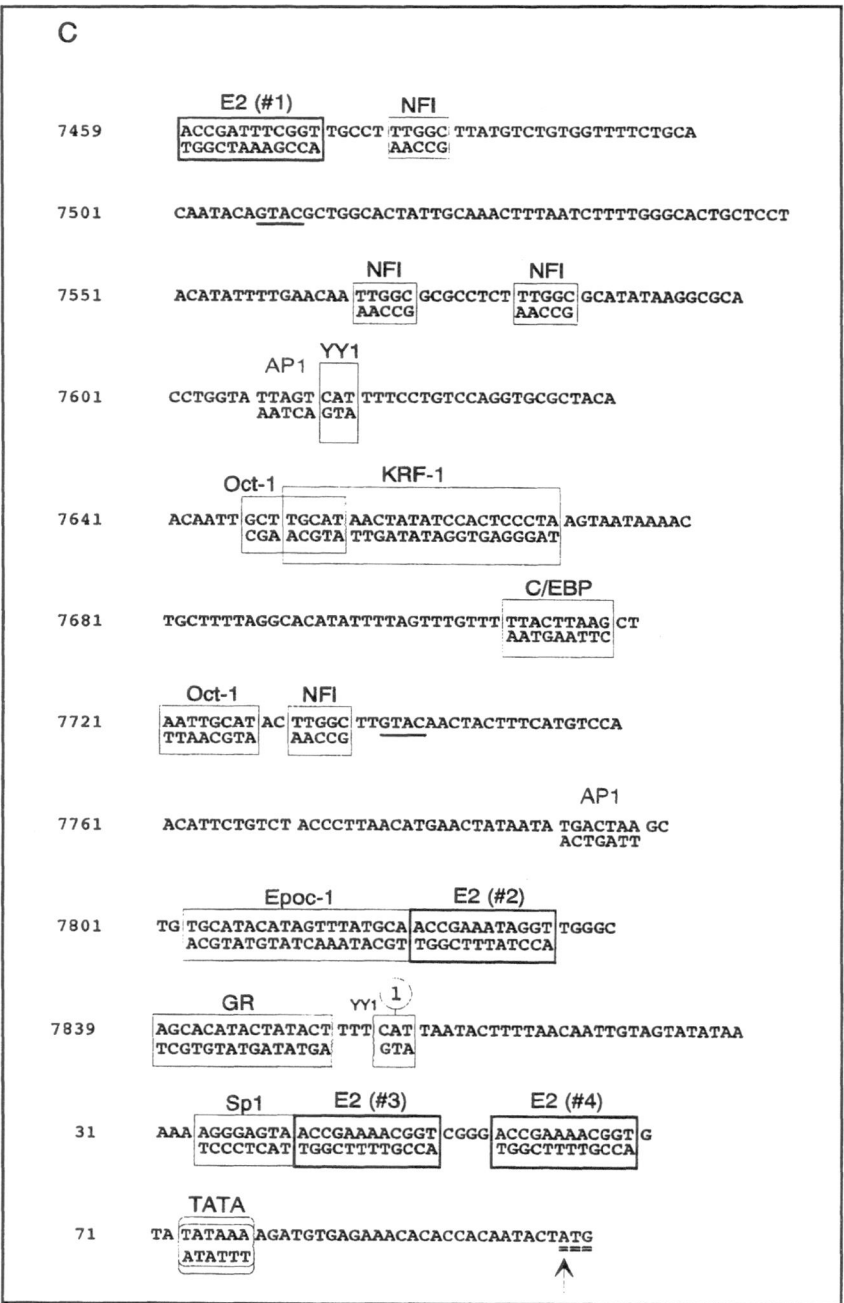

Fig. 2.3C (see legend, opposite page).

protein E1. These various *cis*-elements are located close together or even overlapping and thus provide efficient means for a coordinated regulation of viral DNA replication and transcription.

The central segment of the URR has constitutive enhancer activity preferentially in epithelial cells and is only dependent on cellular factors. This has been shown in particular for HPV16 and HPV 18.[13-16] This indicates that the constitutive enhancer may be important for the transcription of the early viral genes in the initial phase of infection. Furthermore, the cell type specificity of the constitutive enhancer may determine, at least in part, the epithelial cell tropism of the papillomaviruses. Major efforts have been made in order to identify the regulatory *cis*-elements and the transcription factors binding to them, with the ultimate goal to understand the host range restriction of HPVs to epithelial cells and the differentiation-dependent modulation of HPV oncogene expression. Up to now, binding sites for various cellular transcription factors including AP1, NF1, Oct-1, TEF-1, TEF-2, YY1 and the steroid hormone receptors, have been identified in the central segments of the HPV16 and HPV18 URR. None of the various transcription factors, however, is restricted to epithelial cells. Most of

Fig. 2.3. Anatomy of the URR of HPV16 and HPV18.

(A, page 32) Overview indicating the three distinct portions of the URR: the 5'-segment, the central segment with the constitutive enhancer, and the promoter-containing 3'-segment. The positions of the E2 binding sites #1 - #4 are indicated. Binding sites #2 and #3 serve as landmarks for the segmentation of the URR (ref. 21). The nucleotide positions at the end of ORF L1 and of the first ATG codon of ORF E6 are also indicated.

(B and C) Transcription factor binding sites in the central and 3'-segments of the HPV16 URR (B, page 33) and the HPV18 URR (C, opposite). Binding sites are marked by boxes with the names of the transcription factors given above. For the YY1 binding sites, only the central CAT sequence is boxed. The TATA box is indicated by a double-lined frame. The recognition sites for the restriction enzymes EcoRI and DraI are underlined in the HPV16 sequence, and the RsaI cleavage sites in the HPV18 sequence. The nucleotide sequence of the upper DNA strand is given continuously, whereas the nucleotide sequence of the complementary (lower) DNA strand is indicated only in the transcription factor binding sites. Nucleotide position no. 1 is marked by a circled number. The major mRNA start sites from the P_{97} and P_{105} promoters are indicated by arrows. The ATG codons of ORF E6 are doubly underlined.

them are rather ubiquitous factors. Thus the very simple concept that the epithelial cell specificity of the central URR enhancer is due to a transcription factor exclusively present in epithelial cells can be ruled out.

The 5'-segment of the URR contains the signals for termination and polyadenylation of the late gene transcripts. Footprint analyses have indicated that cellular proteins bind to different sites within the 5'-segment of the HPV18 URR.[15] The 5'-segment has no autonomous enhancer activity, but contributes to some extent to the enhancer activity of the complete URR,[15] and includes an enhancer element with a binding site for the transcription factor Sp1.[17] It is still not clear, however, which role the 5'-segment of the URR plays in the regulation of E6 and E7 transcription.

2.3.3. Viral factors

2.3.3.1. The E2 protein

The early HPV proteins are encoded by the genes E6, E7, E1 and E2. They all have the capacity to modulate transcription by either direct or indirect means. Perhaps of major importance for transcriptional regulation at the E6 promoter is the E2 protein. First shown for BPV1, the E2 protein can function as a transcriptional activator and binds as a dimer to palindromic DNA sequences with the consensus sequence $ACCN_6GGT$ (for review see refs. 18, 19). Interaction of the E2 protein with a single E2 binding site only weakly activates transcription, whereas two or multiple E2 binding sites function cooperatively to yield high-level activation. Altogether 12 E2 binding sites are located in the BPV1 URR. Four of them are located upstream of the P_{89} promoter (i.e., the BPV1 E6 promoter) and constitute the major E2-responsive enhancer. The E2 protein has a modular structure with a transcription activation domain at the N-terminus and a domain for DNA-binding and dimerization at the C-terminal part. The two functional domains are separated by a central hinge region which is not conserved between the different papillomaviruses. The BPV1 E2 gene does not only encode the full-length E2 transactivator protein (E2TA), but also two shorter variants called E2TR and E8/E2. The short E2 proteins lack the N-terminal transactivation domain, but retain the C-terminal dimerization and DNA-binding domain. The short E2 proteins interfere with the E2TA transactivator by competing for binding to the E2 palindromes and by forming inactive heterodimers.[20]

Nucleotide sequence comparison has revealed as a characteristic conserved feature that the URRs of the genital HPV types contain four E2 binding sites (numbered 1 to 4 in the 5' to 3' direction of the URR, see Fig. 2.3) with a conserved spacing between the binding sites #2, #3 and #4.[21] The E2 binding site #1 marks the boundary between the 5'-segment and the central segment, site #2 defines the boundary between the central and the 3'-segment, and sites #3 and #4 form a tandem repeat located in close proximity to the TATA box of the E6 promoter.

Analyses of the effects of the E2 proteins on the HPV E6 promoter and on expression of the E6 and E7 oncogenes have yielded apparently conflicting results. On the one hand, it has been reported that the E2 proteins act as repressors of the HPV E6 promoter,[22-25] and that mutations in the E2 ORF increase the frequency of immortalization by the E6 and E7 genes.[26] Due to the close and even overlapping arrangement of transcription factor binding sites in the 3'-segment of the URR, it was deduced that displacement of the cellular factors Sp1 and TFIID by E2 cause the repressive effects.[27] On the other hand, the inability of an HPV16 variant to immortalize cervical keratinocytes has been mapped to a mutation in the E2 gene resulting in a premature stop codon.[28] Furthermore, it was shown in cotransfection experiments performed in cervical keratinocytes that the E2 proteins of HPV16 and HPV18 activate the E6 promoter.[29] Under identical conditions, the BPV1 E2 protein showed repressor activity. Repression by HPV16 E2 was observed when the E2 gene was overexpressed.[29]

In HPV16-immortalized keratinocytes, alternatively spliced mRNAs with the potential to encode a truncated E2 protein are produced. The short E2 protein lacks the N-terminal *trans*-activation domain and thus structurally resembles the short E2 proteins of BPV1. Furthermore, it interferes with the activating properties of the full-length E2 protein.[29]

The contrasting conclusions reached by the different studies about the effects of E2 on the HPV16 or HPV18 URR may be due, at least in part, to the fact that the experiments were performed in different cell types using different E2 expression constructs and enhancer-promoter constructs. The results may not be mutually exclusive but rather indicate that the regulation of transcription of the E6 and E7 oncogenes by E2 is very complex and that the E2 protein plays a dual role as an activator and repressor. The expression levels of E2, the ratio of full-length to short E2 proteins, the presence of cellular transcription factors which cooperate with E2, the differentiation state of the host cell and other parameters may all influence the actual regulatory activity of E2.[20]

2.3.3.2. The E6 protein

One of the major activities of the E6 oncoproteins of the "high risk" HPV types is to interfere with the normal functions of the tumor suppressor protein p53 (see chapter 3). Due to this interaction, E6 can neutralize both the activator and the repressor function of p53 during the cell cycle. Furthermore, the E6 protein can activate or repress transcription from heterologous promoters by p53-independent mechanisms.[31,32]

In the context of this chapter, it is interesting whether the E6 protein can also influence transcription from the URR. It has been shown that the E6 protein of HPV16 activates transcription from several minimal TATA-containing promoters of viral origin, including the HPV16 P_{97} promoter.[33] No E6-responsive enhancer, however, could be detected in the HPV16 URR. These results indicated that the E6 protein might function as a coactivator interacting with general transcription factors such as the TATA-box binding protein, TBP, or others.[33] By using the E6 protein of BPV1, it has been demonstrated that the E6 protein contains a transcription activation domain which coincides with the zinc fingers.[34] The capacity to activate certain promoters like the adenovirus E2 promoter is shared by the E6 proteins of the "high risk" and "low risk" HPV types which differ in their ability to cooperate with E7 in transformation.[35] Thus one may speculate that transcription activation is one of the basic functions of the E6 proteins necessary for the regulation of the viral life cycle. It remains to be clarified, however, whether the transcription factor properties of E6 play a role mainly in order to modulate the expression of certain cellular genes or also in order to regulate the transcriptional activity of the URR.

2.3.3.3. The E7 protein

One of the prominent activities of the E7 oncoprotein of the "high risk" genital HPV types is the complex formation with the cellular tumor suppressor protein pRB (see chapter 4). The binding of E7 releases the transcription factor E2F from complexes with pRB, and the free E2F can bind to its specific *cis*-elements and activate transcription. Many E2F-regulated target genes encode proteins which are necessary for cellular DNA replication and which have to be induced for progression from the G1 phase into the S phase of the cell cycle. By its influence on E2F, the E7 oncoprotein can exert pronounced effects on transcription initiated at E2F-responsive promoters. The papillomavirus URRs, however, do not contain E2F binding sites and the E6 promoter is not E2F-responsive.

The E7 protein can activate transcription from adenovirus early gene promoters also by pRB-independent pathways (see also chapter 4).[36] This may relate to the capacity of the E7 protein to alter the DNA binding properties of transcription factors, as has been shown for the factors ATF and Oct-1.[37] Furthermore, the E7 protein not only binds to pRB and the pRB-related protein p107, but also to other cellular proteins involved in the regulation of transcription. These include the TATA-box binding protein TBP and the TBP-associated factor 110, TAF-110.[38,39] In addition, the E7 protein also interacts with different members of the AP1 family of transcription factors and increases the transcription activation by c-Jun (see chapter 4).[40]

The E7 protein is certainly an important factor to create an intracellular environment suitable for viral DNA replication. Despite the various interactions of E7 with cellular transcription factors it is not yet clear which role the E7 protein plays in the regulation of transcription initiated at the URR E6 promoter. It seems likely, however, that the multifunctional regulator E7 also regulates the transcription of its own gene at least in some phases of the viral life cycle.

2.3.4. Cellular factors

In order to identify the cellular transcription factors interacting with *cis*-elements in the URR of HPV16 and HPV18 and to analyze their effects on the activity of the E6 promoter, several experimental approaches have been employed. Among them, computer-based analysis of the URR nucleotide sequences has been used to identify putative transcription factor binding sites (for review, see ref. 21). The regions of the URR bound by nuclear proteins were identified by footprinting experiments.[15,41,42] Segments of the URRs linked to a heterologous promoter and a reporter gene were analyzed in transient cotransfection assays for their responsiveness to cotransfected cellular transcription factor genes. Mutations were introduced into individual *cis*-elements by site-directed mutagenesis and their effects on the E6 promoter activity in the context of the complete URR were determined in transient transfection analyzes.[43] All these studies showed that particularly the central segment with the constitutive enhancer and the 3'-segment with the E6 promoter contain numerous binding sites for various cellular transcription factors[44] (see Fig. 2.3B for the HPV16 URR and Fig. 2.3C for the HPV18 URR).

2.3.4.1. AP1

AP1 is a dimeric protein complex formed by different members of the jun (c-jun, junB, junD) and fos (c-fos, fosB, fra1, fra2) protooncogene families (for review, see ref. 45). AP1 activates or represses transcription by binding to DNA sequences related to the consensus sequence TGANTCA. The diversity of the family members and the ability of the jun proteins to form either Jun-Jun homodimers or Jun-Fos heterodimers creates a high variability of possible combinations which may have different biological activities. AP1 is the downstream target of intracellular signaling pathways regulated by protein kinase C (PKC) and is regulated by the intracellular redox state.[46,47]

The HPV16 and HPV18 URR contain three and two AP1 binding sites, respectively. For HPV18, it has been demonstrated that both AP1 binding sites are essential for the activity of the E6 promoter since the mutation of either one of them is sufficient to severely reduce the transcription rate.[43,48] In cultured human keratinocytes, JunB has been identified as the predominant Jun component in the AP1 complex binding to the AP1 sites in the HPV18 URR.[48]

It seems probable that AP1 is one of the key players determining the cell type specificity and the differentiation dependence of the transcription activity of the URR.[49] It has been shown that expression of the *jun* and *fos* genes changes during keratinocyte differentiation and that the DNA binding activity of AP1 is increased in the differentiating keratinocytes.[46] Both effects are induced by activation of PKC which seems to mediate the calcium signal for differentiation.[46] Thus changes in the composition of the AP1 complexes during host cell differentiation may affect the rate of transcription initiated at the E6 promoter. In agreement with this assumption is the recent observation that treatment of cells with antioxidants induces alterations in the components of the AP1 complexes which is accompanied by suppression of E6 and E7 oncogene transcription (ref. 50, see also below). In view of the essential role of AP1 in the regulation of HPV transcription, the question is intriguing whether de-regulation of *jun* or *fos* gene expression might play a role in cervical carcinogenesis. The answer is not yet known, but at least amplification and overexpression of the *junB* gene have been reported for two cervical cancer cell lines.[51]

2.3.4.2. YY1

The transcription factor YY1 regulates transcription in a context-dependent manner either as an activator or a repressor, and is a transcription regulator of many viral and cellular genes.[52] YY1 binding sites have been functionally identified in the URR of HPV18[53] and HPV16 URR,[54] and computer-based binding site searches have found potential YY1 binding sites in the URRs of almost all HPV types.[21]

For HPV18, YY1 has been implicated to be an important determinant for the cell-type-specific activity of the URR in HeLa cells. Activation of the HPV18 E6 (P_{105}) promoter requires the binding of YY1 to the promoter-proximal YY1 binding site and the formation of a complex between YY1 and the transcription factor C/EBPβ (CCAAT/enhancer binding protein).[55,56] The YY1-C/EBPβ complex binds to the so-called "switch region" which is located upstream of the YY1 binding site and contains a C/EBP- binding site. YY1 changes into a repressor of the P_{105} promoter when the switch region is deleted or mutated or when only the 3'-segment of the URR is tested.[53,55,56] The analysis of other cell lines revealed that YY1 is either a repressor of the HPV18 URR or has no influence on the transcription activity.[57] Thus the synergistic activation by YY1 and the C/EBP-YY1 complex seems to be an exceptional feature of HeLa cells.

For the HPV16 URR, YY1 seems to function predominantly as a repressor. Interestingly, the extrachromosomal HPV16 DNAs of some cervical carcinomas were found to harbor mutations or deletions of the promoter-proximal YY1 binding sites which abolished the repression by YY1.[25,54] From 24 YY1 binding sites in the HPV16 URR, five are located in the 3'-part of the central enhancer segment in the vicinity of an AP1 binding site. These five sites are required for YY1-mediated repression which seems to occur by quenching the activating activity of AP1.[58] At least some evidence has been provided suggesting that the repression of AP1 function by YY1 does not occur by preventing AP1 from binding to its specific DNA site, but rather that binding of YY1 to the coactivator CPB (CREB-binding protein) may be involved.[58]

From the data assembled so far, the picture seems to emerge that interference between the activator AP1 and the repressor YY1 plays an important role in the regulation of URR transcription activity. As shown for HeLa cells, the repressor YY1 can efficiently turn into an activator depending on the presence of other transcription factors with which YY1 can interact. Escape from YY1-mediated repression by mutations of the YY1 binding sites seems to be one mechanism by which the expression of the E6 and E7 genes of the oncogenic HPV types (at least HPV16) can be increased.

2.3.4.3. C/EBP

The C/EBP- (CCAAT/enhancer binding protein) factors form a family of sequence-specific DNA binding proteins which contain a basic zipper motif for DNA binding and protein dimerization. The C/EBPs are involved in the regulation of cell differentiation, they activate the expression of several cytokine genes and bind to the promoter regions of DNA viruses. Several lines of evidence have accumulated indicating that C/EBPs are regulators of HPV gene expression. C/EBPβ, first identified as the nuclear factor for expression of interleukin 6 and therefore also called NF-IL6, has been shown to bind to several sites in the HPV16 URR and to repress the P_{97} promoter.[59,60] Since the C/EBP binding sites overlap with binding sites for NF-1 and AP1, it has been assumed that competition with activator proteins for binding to the URR might cause the repressive effect.[60] C/EBPβ has also been found to bind to the HPV11 URR and to repress HPV11 transcription and DNA replication in cultured foreskin keratinocytes.[61] Furthermore, C/EBPβ binds to the switch region of the HPV18 URR and forms a complex with YY1 which is necessary for the YY1-mediated activation of the P_{105} promoter.[55]

C/EBP expression is modulated by many factors which are involved in the regulation of epithelial cell growth and differentiation, e.g., by epidermal growth factor, calcium and retinoids (see ref. 61). Furthermore, C/EBP and AP1 can either cooperate or interfere in the regulation of target gene transcription.[62,63] These and other properties make C/EBPβ and potentially the other members of the C/EBP family highly interesting candidates that participate in the complex network of regulation of E6 and E7 oncogene transcription.

2.3.4.4. Sp1

In the URRs of most HPVs, a binding site for the transcription factor Sp1 can be found in close vicinity to the TATA box and the two E2 binding sites in the 3'-segment.[21] Sp1 is a zinc finger protein that binds to GC box elements which are present in the promoter regions of many cellular and viral genes. The promoter-proximal Sp1 site is essential for the transcriptional activity of the URR.[43,64] Since the 3'-segment lacks intrinsic enhancer activity, binding of Sp1 to the promoter-proximal site seems to be essential mainly for the communication of other transcription factors bound to upstream URR-elements with the basal transcription machinery. Due to the close proximity of their binding sites, the viral E2 protein can interfere with Sp1 binding to the URR.[27,65] This interference may be important for the coordinate regulation of viral gene expression and DNA replication.

Mutation of the promoter-proximal Sp1 site in the HPV18 URR led to a much stronger reduction of the P_{105} promoter activity in epithelial cells than in nonepithelial cells.[66] This result indicates that Sp1 may be one of the factors determining the epithelial cell preference of the HPV URR.

An additional binding site for Sp1 has been identified in the 5'-segment of the HPV18 URR which, however, is probably not involved in the regulation of E6 promoter activity.[17]

2.3.4.5. Nuclear receptors

The nuclear receptors comprise a large superfamily of ligand-dependent transcription factors which bind to specific hormone-responsive DNA elements located in the promoter regions of target genes. The nuclear receptor superfamily includes the steroid hormone receptors and the thyroid/retinoid receptors.[67] Whereas the steroid hormone receptors activate transcription by binding to

DNA as homodimers, the thyroid and retinoid receptors predominantly bind to DNA as heterodimers (for review, see 68 and 69).

DNA binding sites for both glucocorticoid and progesteron receptors have been characterized in the URRs of HPV16, HPV18 and HPV11.[43,70-72] The single glucocorticoid-responsive element (GRE) in the HPV18 URR is located in the 3'-segment between the AP1 and Sp1 binding sites. The HPV16 URR contains three GREs which are located in the 5'-segment and the central enhancer. The GREs confer steroid hormone responsiveness to the homologous E6 promoter or to heterologous promoters. Mutation of the GRE in the context of the entire HPV18 URR results in a complete loss of dexamethasone-dependent inducibility. Whereas a double point mutation within the GRE led to a consistent increase of the basal activity of the P_{105} promoter,[43] a single point mutation decreased the basal activity.[72] The contrasting effects of the different mutations remain to be explained.

The activation of the URR by steroid hormones as tested in transient transfection assays with reporter gene constructs correlates well with the observation that steroid hormone treatment results in an increased transformation of target cells by the HPV E6 and E7 oncogenes.[73]

In contrast to the GREs, no retinoic acid-responsive elements have yet been identified in the URRs. Nevertheless, ligand-activated retinoid receptors reduce the transcription rate at the E6 promoter of HPV18 and HPV16.[74] The mechanism has not been clarified, but an interference with the activator function of AP1 may be possible. Antagonistic interactions between the retinoid receptors and AP1 are involved in the regulation of target gene transcription and are important for cell differentiation and proliferation (for review, see 75). Mutual interference with AP1 has also been demonstrated for the steroid hormone receptors.[68]

Altogether the studies have shown that different members of the nuclear receptor family participate in the regulation of E6 and E7 gene transcription. The steroid hormone receptors act directly by binding to GREs in the URR, whereas the retinoid receptors more likely act indirectly. Since the steroid hormones and retinoic

acid are essential regulators of the proliferation and differentiation of cervical epithelial cells, it seems likely that they are important for the regulation of HPV gene expression in vivo.

2.3.4.6. KRF-1

A keratinocyte-specific factor, KRF-1, has been found to bind
to a DNA element in the constitutive enhancer of the HPV18 URR
and to interact with AP1 in the activation of transcription.[76] Mutation of the KRF-1 binding site led to a strong reduction of URR
activity when tested in cells of the spontaneously immortalized
keratinocyte line HaCaT or in primary keratinocytes, whereas in
HeLa or C33A cervical carcinoma cells only a slightly negative effect was observed.[43] These data suggested that carcinoma cells may
harbor alterations which can compensate for the loss of KRF-1
activity. The KRF-1 factor does not appear to interact with URR
sequences of other genital HPV types like HPV16 and HPV11,[76]
indicating that related HPV types have different requirements for
cellular transcription factors in order to activate the early promoter.
The identity of the KRF-1 factor has not yet been unraveled.

2.3.4.7. POU domain proteins

The POU domain is a conserved bi-partite motif for sequence-
specific DNA binding and protein-protein interactions present in
a variety of transcription factors which are involved in the regulation of development and differentiation (for review, see ref. 77).
The POU domain proteins bind to the canonical octamer DNA
sequence ATGCAAAT or to degenerate sequence motifs and are
therefore also named octamer-binding or Oct proteins.

The Oct-1 protein was first described to repress the HPV18
URR, an effect which is independent of the DNA binding of Oct-1
but requires Oct-1 overexpression.[78] Two sequence-aberrant Oct-1
binding sites have been identified in the constitutive enhancer of
the HPV18 URR. Mutation of the Oct-1 binding site located in the
3'-part of the enhancer in the context of the complete URR did not
result in a strong modulation of E6 promoter activity in HeLa or
HaCaT cells indicating that, at least under the experimental con-

ditions used, the transcription rate from the E6 promoter is independent of the Oct-1 element.[43]

In the constitutive enhancers of HPV16, HPV18 and other genital HPVs, a conserved composite regulatory element has been identified which consists of a degenerate octamer motif (previously named NFA site, ref. 79) and a nonpalindromic NFI site separated by 2 bp.[80] Mutation of either one of the binding sites resulted in a decrease of HPV16 enhancer activity which could not be intensified by the simultaneous mutation of both sites. Protein binding studies suggested that the Oct-1 protein has the function to tether NFI to the composite element.[80] The octamer motif of the composite element described by O'Connor and Bernhard corresponds to the Oct-1 binding site in the 3'-part of the enhancer.[43,80] The two studies revealed somewhat different effects of the mutated Oct-1 binding sites on promoter activity which are likely to be due to differences in the experimental designs.

Furthermore, Oct-1 has been shown to bind to another site in the central enhancer of the HPV16 URR which is located upstream of the composite element. This Oct-1 site overlaps with the binding site for the factor PEF-1 (papillomavirus enhancer binding factor 1, see ref. 81). Mutation of the PEF-1 binding site reduced and mutation of the Oct-1 binding site enhanced the URR activity.[81] Other factors binding to this region are NF-I and a novel methylation-sensitive transcription factor MSPF.[41,82]

Recently a newly identified POU domain transcription factor, called Epoc-1 in mouse and skn-1a in rat, has been shown to bind to the HPV18 URR and to activate the E6 promoter of HPV18 and HPV16.[83] The Epoc-1/skn-1a gene is expressed predominantly in the skin, where transcripts are detectable in the differentiating suprabasal cells, but not in the basal cells (for all references, see ref. 83). One Epoc-1 binding site has been localized in the 3'-segment of the HPV18 URR in close proximity to the AP1 binding site. Epoc-1 binding sites in the HPV16 URR have not yet been determined.

The expression profile of Epoc-1 in the squamous epithelium of the skin and its ability to transcriptionally activate the HPV URR strongly suggests that Epoc-1 is involved in the differentiation-dependent regulation of HPV gene expression. Since transcription factors also regulate DNA replication (for a review see ref. 84), Epoc-1 may also play a role in vegetative HPV DNA replication which occurs only in the highly differentiated cells. Interestingly, Epoc-1 does not seem to play a role for transcription of the E6 and E7 oncogenes in cervical carcinoma cells suggesting that the carcinoma cells have escaped the requirement of Epoc-1 for high-level transcription of the HPV oncogenes.[83]

Altogether the studies have provided evidence that POU domain proteins are involved in the regulation of E6 and E7 oncogene expression. However, many questions have still to be answered. Computer searches find degenerate octamer motifs in the HPV URRs in addition to those analyzed so far.[21] Their functional role remains to be determined. Furthermore, the POU protein family is large and includes additional members expressed in keratinocytes, like Oct-6,[85] and probably also members not yet identified. It remains to be clarified whether other Oct proteins also play a role in the regulation of E6 promoter activity.

2.3.4.8. NF I

The NFI (nuclear factor I) family includes several transcription factors which are encoded by different genes (NFI-A, -B-, -C and -X) and regulate the transcription of many cellular and viral genes. NFI is also important for the initiation of adenovirus DNA replication. NFI proteins bind to a consensus sequence motif 5'-TGG(A/C)N$_5$GCCAA with a DNA binding domain containing conserved cysteine residues. In epithelial cells mainly the NFI-C gene is expressed whereas NFI-X gene products predominate in fibroblasts (for references see ref. 86).

Computer-based sequence analyses and protein binding studies have revealed that the papillomavirus URRs contain several NFI binding sites mainly in the constitutive enhancer region.[42,79] These binding sites consist only of the half-site sequence TTGGC and bind NFI protein with low affinity. The NFI binding sites of the HPV16 enhancer have been shown to be important for conferring epithelial cell-specific activity to a heterologous promoter.[79,87,88] Mutation of the NFI sites of the HPV18 enhancer in the context of the complete URR, however, generated only a slight decrease of promoter activity.[43] These at the first sight contradictory results might be explained by differences in the experimental design, in particular by the usage of different reporter constructs. An alternative but less likely possibility is that the NFI sites are functionally important only in HPV16 but not in HPV18. As already mentioned, one of the NFI binding sites is part of the conserved composite regulatory element which seems to trigger an interaction between NFI and Oct-1.[80] A decrease of the HPV16 enhancer activity in epithelial cells could be achieved by overexpression of the NFI-X1 gene which is usually expressed in fibroblasts.[86] This finding provides another example for the important phenomenon that the members of a transcription factor family can exert different influences on a target gene promoter.

2.3.4.9. TEF

The transcriptional enhancer factors 1 and 2 (TEF-1, TEF-2) were originally isolated due to their specific binding to the SV40 enhancer. TEF-1 is an abundant protein in keratinocytes, but it is also present in other cell types with the exception of lymphocytes. TEF-1 has been shown to bind to four sites in the HPV16 enhancer and to be an activator of the P_{97} promoter which requires the interaction with a cell-specific coactivator protein.[89] TEF-1 and its coactivator are present in cell types in which the HPV16 URR is inactive, demonstrating that TEF-1 is insufficient to trigger the epithelial cell specificity of the HPV URR.[89] TEF-1 binds to sequences which contain a central CAT motif. One of the TEF-1 binding sites in the HPV16 URR is also recognized by TEF-2 which was previously named PVF.[79,87] The binding sequences for TEF-1 are similar to those for TEF-2 and YY1 thus leaving the possibility that a recognition site for TEF-1 is also contacted by TEF-2 or YY1 and vice versa. The role of TEF proteins for the activity of the HPV18 URR has not yet been examined.

2.4. THE CONSEQUENCES OF INTEGRATION
ON VIRAL GENE EXPRESSION

Recombination between the viral DNA with host cellular sequences represents presumably an essential step in HPV-linked carcinogenesis. In premalignant cells, HPV16/18 commonly persist as autonomously replicating episomes in a defined copy number,[90] whereas the viral DNA is found to be integrated in the vast majority of cervical carcinomas and cell lines derived therefrom.[91,92] Southern blot analyses of different specimens indicate that integration seems to occur very early during the multi-step process of cervical cancer.[93,94] Integration often leads to a disruption of the early coding region located between the 5'-end of the E1 and the 3'-end of the E2 ORF (see refs. 95-96, see also Fig. 2.4 as an overview). While the URR as well as the E6/E7 ORFs were consistently found to be intact in malignant lesions[95,97] loss of E2 function may provide a selective advantage toward malignant progression, since E2 negatively interferes with the efficiency of E6/E7 expression via binding to the cognate E2 binding motif within the proximal end of the viral URR.[23,24] Consequently, viral integration relieves the *trans*-repressing function of the E2 protein on its own URR, which in turn favors increased expression of HPV specific oncoproteins. Functionally equivalent to the situation found in vivo are in vitro DNA-mediated gene transfer experiments using full-length HPV16 or HPV18 DNA harboring a point mutation or deletion within the E2 ORF. Transfection of such mutants on primary human keratinocytes leads in both cases to enhanced frequency of immortalization,[26,98] which again supports the notion that E2 can principally be considered as a negative regulator during the viral transformation process.

Another consequence of viral integration concerns the mode of termination of HPV transcription and the regulation of the cytoplasmic half-life of the virus-specific mRNAs. Various studies have shown that integration of the HPV DNA into the host cellular genome generates chimeric transcripts where the viral 5'-part containing the E6 and E7 ORFs is fused to cotranscribed downstream cellular sequences.[5,96,99,100] Such an arrangement obviously provides an additional benefit for the high-risk HPV types during the multi-step process of cellular transformation, since the virus-specific A+U-rich RNA labilization signal, located nearby the 3'-end of the early coding region, becomes deleted.[101] This could also account for accelerated progression to cervical cancer, because an increased half-life of the fusion transcripts guarantees elevated accumulation of the viral oncoproteins and probably a dose-responsive effect on malignant transformation.[102]

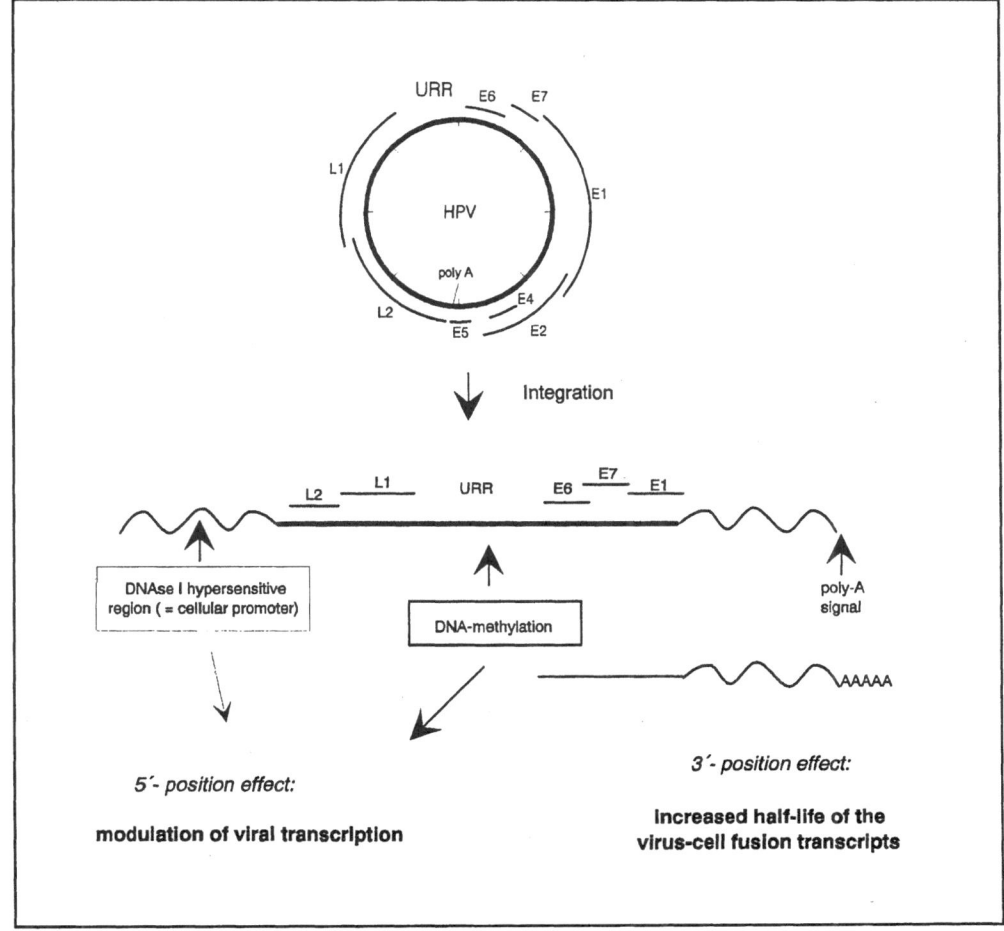

Fig. 2.4. Schematic overview about the consequences of viral integration on HPV specific gene expression.

The availability and utilization of a cellular polyadenylation signal to terminate viral transcription is also indicative of the existence of a corresponding cellular promoter in a further upstream chromosomal region. One experimental way to scan viral integration loci for the existence of additional cellular regulatory elements without the knowledge of the surrounding primary nucleotide sequences is the mapping analysis of so-called DNAse I hypersensitive sites. Nuclease hypersensitive sites represent not only known hallmarks for regulatory elements at the chromatin level,[103] but also seem to be preferred targets for retroviral insertions.[104] By investigating the nucleosomal organization of the HPV18 integration locus in the cervical carcinoma cell line HeLa, it could be demonstrated that HPV18 is integrated in close proximity of a cellular promoter which cooperates with the viral URR in generating the characteristic HPV18 transcription pattern.[105] The utilization of an additional heterologous cellular promoter also explains previous cDNA sequencing data showing that one abundant viral transcript, which contains the genetic information for the E6/E7 proteins, also encodes parts of the HPV18 URR, including the two TATA boxes, normally used to initiate early transcription.[106] Monitoring the viral integration locus in the cervical carcinoma cell line SiHa, HPV16 was also found to be integrated in the vicinity of two prominent DNAse I hypersensitive regions.[107] It is therefore reasonable to assume that a decondensed chromatin structure within the host genome not only favors the integration of retroviruses,[104] but also the recombination between DNA tumor viruses and cellular DNA.

Position effects are obviously not merely coincidental, but can indeed affect the transcriptional activity of the regulatory region and viral oncogene expression. For instance, depending on the chromosomal integration site in different cervical carcinoma cells, adjacent *cis*-acting cellular regulatory elements seem to modify the glucocorticoid regulated response of the HPV18 URR[70] in a hierarchical manner, since beside upregulation of viral transcription in the cervical carcinoma cell line C4-1, E6/E7 expression was also found to be down-regulated in SW756 cells or even response-refractory in HeLa cells upon dexamethasone treatment, despite of the presence of a functional hormone receptor and a corresponding binding site within the viral control region.[108]

Expression of integrated viral DNA cannot only be controlled by cellular position effects as outlined above, but also by epigenetic alterations such as de novo DNA methylation and nucleosomal packageing.[109-111] These mechanisms seem to play a considerable role in HPV-linked carcinogenesis, since recombination of the conserved URR-E6/E7 transcription cassette with the host genome subjects the viral DNA to cellular modification processes which normally affect the expression of the unoccupied gene or chromosomal domain.

It has been demonstrated that in vitro methylation of the upstream regulatory region of either HPV16 or HPV18 results in a selective downregulation of the transcriptional activity, a phenomenon, which can be attributed to a cluster of 5'-GpCpGpC-3'sites within the virus specific enhancer element.[82,112] Moreover, de novo methylation of CpG residues within the E2 binding site also abrogates E2 binding, indicating that the balanced modulation of E6/E7 expression via the antagonistic activity of the E2 protein cannot only be destroyed by viral integration, but also by epigenetic modification processes.[113] In vivo competition experiments have revealed that the suppression effect is not simply mediated by steric hindrance of the methyl-residues and the exclusion of transcription factors from the DNA template, but rather indirectly by an enhanced affinity of titratable methyl-DNA binding proteins.[114,115] This particular class of proteins seems to be also involved in the nucleosomal assembly process of both methylated cellular and viral DNA.[112,116,117] By monitoring the chromatin structure of highly methylated HPV16 DNA in the cervical carcinoma cell line CaSki, it could be demonstrated that most of the tandemly integrated multi-copy DNA is organized in a nucleosomal fashion characteristic for transcriptionally inactive chromatin.[112] This may explain the observation, that there is a rather low level of E6/E7 expression despite the presence of more than 600 copies of HPV16 in this specific cell line.

Hence, DNA methylation is not only important for controlling host cell gene expression (for review, see ref. 118), but also for the transcriptional regulation of human papillomaviruses. De novo methylation can principally be considered a host defense mechanism directed against foreign integrated DNA.[110] Exactly such epigenetic modification processes may account for the generation of a kind of viral latency of HPV genomes similar to that described recently for the HIV proviral DNA or for the cottontail rabbit papillomavirus in transgenic rabbits.[119,120] Consistent with this interpretation is also the finding that in vitro methylated bovine papillomavirus type 1 DNA shows a reduced potential to transform heterologous rodent cells.[121] Because progression towards malignancy is often accompanied by regional hypomethylation of the genomic DNA (for review, see ref. 122), it is conceivable that methylated HPV genomes become additionally re-activated which in turn may accelerate malignant conversion by elevated E6/E7 gene expression. In fact, recent epidemiological data have shown that the lack of folic acid, which is an essential component of our nutrition, increases the incidence of cervical cancer.[123] This is an important aspect, since the lack of folic acid seems to reduce the level of methyl groups in eukaryotic cells.[124]

2.5. THE ROLE OF THE GENETIC BACKGROUND: TRANSCRIPTIONAL REGULATION OF VIRAL GENE EXPRESSION IN NONMALIGNANT AND MALIGNANT CELLS

In vitro transfection studies have shown that the E6 and E7 genes of HPV16 or HPV18 or other "high risk" HPV types are capable of immortalizing primary human keratinocytes and fibroblasts.[125-127] In vivo experiments clearly indicate that such cells are not tumorigenic in immunocompromised animals.[127,128] Only their long term in vitro cultivation or the subsequent introduction of additional oncogenes malignantly transform HPV16/18 immortalized cells, leading in both cases to fast growing tumors in nude mice.[128-132]

These experimental observations are in line with epidemiological data[133] showing that viral transcription in cervical intraepithelial neoplasias is obviously not sufficient to induce malignancy, since there is a long latency period between primary infection and the final clinical manifestation of a tumor.[134,135] Consistent with the multi-step concept of human cancer, it has been postulated that nontumorigenic HPV positive cells still harbor regulatory control mechanisms, which negatively interfere with both viral gene expression and tumorigenicity.[134] It was further hypothesized that the upstream regulatory region of high-risk HPV types represents a major target of such an intracellular control pathway, whose genetic loss is accompanied by uncontrolled viral transcription and ultimately by the progression to a tumorigenic phenotype.[134] Expression of viral oncogenes especially during the initial stage of HPV-linked carcinogenesis may result in a destabilization of the karyotype of the infected host cell. This notion is supported by the findings that long term expression of viral oncogenes such as the SV 40 T-antigen or E7 of HPV 16 induces an accumulation of chromosomal abnormalities, which is finally accompanied by the segregation of genetic information in form of individual alleles.[136,137]

Indeed, it became more and more evident during the last years that the loss of genetic function plays also a pivotal role in the development of cervical cancer. This has been mainly deduced from somatic cell hybridization experiments with cervical carcinoma cells and normal human fibroblasts or keratinocytes.[138] A detailed comparison of restriction-fragment-length-polymorphisms of the DNA from nontumorigenic HeLa-fibroblast hybrids, their tumorigenic revertants and the parental cervical carcinoma cells have provided evidence for a nonrandom loss of chromosome 11.[139,140] Moreover, micro-cell transfer experiments of individual chromosomes directly prove the effect of chromosome 11 on the in vivo growth properties of cervical carcinoma cell lines, since the re-introduction of a normal copy of chromosome 11 either in the HPV18-positive cell line HeLa or the HPV16-containing SiHa cells result in a suppression of the tumorigenic phenotype.[141,142]

Furthermore, an even direct involvement of chromosome 11 on the transcriptional regulation of HPV16 expression was documented by another set of studies. It could be demonstrated that reporter constructs driven by the HPV16-URR were silent in primary human fibroblasts, but highly active in fibroblasts carrying a deletion at the short arm of chromosome 11.[143] This experimental model system provides a functional link between the efficiency of HPV expression and intracellular phosphorylation processes, since cells which harbor this particular deletion on chromosome 11 (Del-11) show an upregulation of the inhibitory subunit (PR55) of the phosphatase 2A (PP2A).[144] Interestingly, elevated expression of the PR55 subunit can also be induced by the overexpression of the SV 40 small t antigen or by treatment with the PP2A inhibitor okadaic acid. In both cases the transcriptional activity of the HPV16-URR can be restored even in primary human fibroblasts.[144]

Somatic cell hybrids made between cervical carcinoma cells and normal human fibroblasts or keratinocytes have also been used to demonstrate that "high risk" HPVs are differentially regulated in cells with altered genetic background. It has been shown that 5-azacytidine, a substance which reactivates the expression of genes silenced by hypermethylation (for review, see ref. 145), is capable of selectively suppressing HPV18 transcription exclusively in non-tumorigenic HeLa-fibroblast-hybrids, while E6/E7 expression was unaffected in tumorigenic segregants or parental HeLa cells.[146] Downregulation of viral oncogene expression is accompanied by cellular growth arrest and can be abrogated by blocking protein synthesis, indicating that (a) repressor(s) is (are) involved in this process. Although the molecular basis of this phenomenon is still not understood, it is reasonable to assume that 5-azacytidine may induce genes, which are formally transcriptionally silenced by de novo methylation probably due to their incompatibility with cell proliferation or the differentiation status of the cells.[117,145] This assumption is supported by recent data, which demonstrate that cell cycle inhibitors such as the cyclin D/cyclin-dependent kinase inhibitor p16 is indeed found to be epigenetically modified in proliferating cells (for review, see ref. 147). While p16 is upregulated in human cells undergoing senescence, the gene is found to be de novo methylated and not expressed in immortalized human keratinocytes.[148]

As outlined above, analysis of negative regulation directly on HPV transcription is hampered by the fact that continuous E6/E7 expression seems to be an essential prerequisite to maintain cell proliferation in vitro and in vivo.[149,150,151] Attempts to down-regulate viral oncogene expression are regularly accompanied by a cessation of cellular growth.[149,150] To circumvent this apparent problem, an in vitro system was developed where putative suppression mechanisms on a viral control region can be investigated in actively growing cells, without any interference with the proliferative phenotype. This was achieved by transfecting a HPV18-URR driven chloramphenicol-acetyl-transferase reporter construct (CAT) into a heterologous HPV16 positive cervical carcinoma cell line (SiHa) under nonselective conditions. While the reporter gene is highly expressed in malignant cells, HPV18-directed CAT expression becomes completely suppressed after fusion with nonmalignant human keratinocytes. Control experiments using tumorigenic SiHa cells as a fusion partner did not show any influence on HPV18-driven CAT expression. These results strongly suggest that nontumorigenic cells harbor *trans*-acting regulatory factors negatively interfering with the HPV18 regulatory region, and that these factors are missing in tumorigenic cervical carcinoma cells.[152]

Cell hybridization experiments demonstrated that both immortalized human keratinocytes and cervical carcinoma cells can be attributed to at least four genetically different complementation groups with regard to the induction of cellular senescence or suppression of the tumorigenic phenotype.[153,154] Interestingly, similar to the aforementioned SiHa-CAT-model system, where the upstream regulatory region of a HPV "high risk" type became extinguished when assayed under nonselective conditions, viral gene expression of one fusion partner could be also suppressed if two malignant HPV16 and HPV18 cell types were used for hybrid formation.[154]

The loss of intracellular control mechanisms acting at the level of transcription initiation seems also to play a substantial role during progression to cervical cancer. The viral URR seems to be a target of negative regulation, since in situ hybridization studies have shown that there is very low transcriptional activity of "high-risk" HPV types in basal cells derived from low grade cervical intraepithelial neoplasias (CIN I). Viral transcription, however, gradually increases in higher graded lesions (CIN III) and is abundant in cervical biopsy specimens.[1,155] Similar observations can also be made in heterotransplantation experiments using nontumorigenic HeLa-fibroblast hybrids.[156] Although the nonmalignant hybrids, their tumorigenic segregants as well as the parental HeLa cells, share similar in vitro growth properties, HPV18 transcription is found to be selectively suppressed only in nontumorigenic cells, a phenomenon, which precedes the cessation of cellular growth.[156] All these data clearly argue for the existence of an intracellular surveillance mechanism, which controls the expression of E6/E7 at the level of initiation of transcription.[134,157]

On the other hand, it should be noted that there is also experimental evidence for an additional cellular control pathway. Using different HPV immortalized human keratinocytes as fusion partners, it could be demonstrated that hybrid cells start to undergo senescence despite on-going viral transcription.[153] This indicates that HPV gene expression is not only controlled at the level of initiation of transcription, but also via post-translational mechanisms by neutralizing the capacity of E6/E7 to induce cell proliferation.[135,158]

Which transcription factors are actually involved in the negative regulation of the viral URR is still unknown. However, as already summarized above, many studies have identified several potential candidates, such as Oct-1, YY1, NF-IL6 (C/EPB) and the retinoic acid receptors.[25,52,54,56,60,74,78,126,159] Furthermore, despite the fact that AP-1 is considered as a key activator of E6/E7 oncogene transcription,[14,43,48,49] recent experiments have shown that modified AP-1 could obviously also be involved in viral gene suppression. This is mainly deduced from experiments showing that treatment of cells with antioxidants induces alterations in the components of AP-1, which is paralleled by a selective downregulation of E6/E7 gene expression on the level of initiation of transcription. Although the c-*fos* gene is also strongly induced by the antioxidant, the corresponding protein is not associated with c-Jun to form the prototype AP-1 complex. Instead, c-Jun becomes predominantly heterodimerized with Jun B and Fra-1, which are both components negatively interfering with the role of AP-1 as a positive regulator.[160,161] The observation that c-Jun preferentially forms heterodimers with Fra-1 under condition where viral gene is suppressed is intriguing. Fra-1 is localized in a chromosomal region (11q13), which is often structurally affected in cervical carcinoma cells.[162,163] This could be of particular relevance with respect to the tumor suppressing function of chromosome 11 and the different regulation of HPV transcription in malignant and nonmalignant cells.[135,141,142,146,152] It also interesting to note that YY1 obviously represses HPV transcription by quenching the activity of AP-1, a transcription factor indispensable for successful gene expression.[58] Which of these transcription factors are actually involved in HPV gene control during the multi-step process to cervical cancer in vivo is presently not clear. However, in situ-hybridization experiments on biopsy specimens from different stages should shed some light in this open question.

REFERENCES

1. Crum CP, Nuovo G, Friedman D et al. Accumulation of RNA homologous to human papillomavirus type 16 open reading frames in genital precancers. J Virol 1988; 62:84-90.
2. Dürst M, Glitz D, Schneider A et al. Human papillomavirus type 16 (HPV16) gene expression and DNA replication in cervical neoplasia: analysis by in situ hybridization. Virology 1992; 189:132-140.
3. Stoler MH, Rhodes CR, Whitbeck A et al. Human papillomavirus type 16 and 18 gene expression in cervical neoplasias. Hum Pathol 1992; 23:117-128.
4. Crum CP, Symbula M, Ward BE. Topography of early HPV16 transcription in high-grade genital precancers. Am J Pathol 1989; 134:1183-1188.
5. Schneider-Gädicke A, Schwarz E. Different human cervical carcinoma cell lines show similar transcription patterns of human papillomavirus type 18 early genes. EMBO J 1986; 5:2285-2292.
6. Sherman L, Alloul N, Golan I et al. Expression and splicing patterns of human papillomavirus type 16 mRNAs in precancerous lesions and

carcinomas of the cervix, in human keratinocytes immortalized by HPV16, and in cell lines established from cervical cancers. Int J Cancer 1992; 50:356-364.

7. Roeder RG. The role of general initiation factors in transcription by RNA polymerase II. Trends Biochem Sci 1996; 21:327-335.

8. Chao DM, Young RA. Activation without a vital ingredient. Nature 1996; 383:119-120.

9. Koleske AJ, Young RA. The RNA polymerase II holoenzyme and its implication for gene regulation. Trends Biochem Sci 1995; 20:113-116.

10. Struhl K. Chromatin structure and RNA polymerase II connection: implications for transcription. Cell 1996; 84:179-182.

11. Goodrich JA, Cutler G, Tjian R. Contacts in context: promoter specificity and macromolecular interactions in transcription. Cell 1996; 84:825-830.

12. Cowell IG. Repression versus activation in the control of gene transcription. Trends Biochem Sci 1994; 19:38-42.

13. Gloss B, Bernhard HU, Seedorf K et al. The upstream regulatory region of the human papillomavirus-16 contains an E2 protein-independent enhancer which is specific for cervical carcinoma cells and regulated by glucocorticoid hormones. EMBO J 1987; 6:3735-3743.

14. Cripe TP, Alderborn A, Anderson RD et al. Transcriptional activation of the human papillomavirus-16 P97 promoter by an 88-nucleotide enhancer containing distinct cell-dependent and AP-1-responsive modules. The New Biologist 1990; 2:450-463.

15. Garcia-Carranca A, Thierry F, Yaniv M. Interplay of viral and cellular proteins along the long control region of human papillomavirus type 18. J Virol 1988; 62:4321-4330.

16. Gius D, Grossman S, Bedell MA et al. Inducible and constitutive enhancer domains in the noncoding region of human papillomavirus type 18. J Virol 1988; 62:665-672.

17. Hoppe-Seyler F, Butz K. A novel cis-stimulatory element maps to the 5' portion of the human papillomavirus type 18 upstream regulatory region and is functionally dependent on a sequence-aberrant Sp1 binding site. J Gen Virol 1993; 74:281-286.

18. Ham J, Dostatni N, Gauthier JM et al. The papillomavirus E2 protein: a factor with many talents. Trends Biochem Sci 1991; 16:440-444.

19. Howley PM. Papillomavirinae: the viruses and their replication. In: Fields BN et al, eds. Virology. 3rd ed. Philadelphia: Lippincott-Raven Publishers, 1996:2045-2076.

20. McBride AA, Romanczuk H, Howley PM. The papillomavirus E2 regulatory proteins. J Biol Chem 1991; 266:18411-18414.

21. O'Connor MJ, Chan S-Y, Bernard H-U. Transcription factor binding sites in the long control region of genital HPVs. In: Myers G et al, eds. Human Papillomaviruses 1995. A Compilation and Analysis of Nucleic Acid and Amino Acid Sequences. Part III. Los Alamos: Los Alamos National Laboratory, 1995:47-57.

22. Thierry F, Yaniv M. The BPV1-E2 trans-acting protein can act either as an activator or a repressor of the HPV18 regulatory region. EMBO J 1987; 6:3391-3397.

23. Bernard BA, Bailly C, Lenoir MC et al. The human papillomavirus type 18 (HPV18) E2 product is a repressor of the HPV18 regulatory region in human keratinocytes. J Virol 1989; 63:4317-4324.

24. Romanczuk H, Thierry F, Howley PM. Mutational analysis of cis elements involved in E2 modulation of human papillomavirus type 16 P_{97} and type 18 P_{105} promoters. J Virol 1990; 64:2849-2859.

25. Dong XP, Stubenrauch F, Beyer-Finkler E et al. Prevalence of deletions of YY-1 binding sites in episomal HPV16 DNA from cervical cancer. Int J Cancer 1994; 58:803-808.

26. Romanczuk H, Howley PM. Disruption of either the E1 or the E2 regulatory gene of human papillomavirus type 16 increases viral immortalization capacity. Proc Natl Acad Sci USA 1992; 89:3159-3163.

27. Tan SH, Leong LE, Walker PA et al. The human papillomavirus type 16 E2 transcription factor binds with low cooperativity to two flanking sites and represses the E6 promoter through displacement of Sp1 and TFIID. J Virol 1994; 68:11-20.

28. Storey A, Greenfield I, Banks L et al. Lack of immortalizing activity of a human papillomavirus type 16 variant DNA with a mutation in the E2 gene isolated from normal human cervical keratinocytes. Oncogene 1992; 7:459-465.

29. Bouvard V, Storey A, Pim D et al. Characterization of the human papillomavirus E2 protein: evidence of trans-activation and trans-repression in cervical keratinocytes. EMBO J 1994; 13:5451-5459.

30. Ushikai M, Lace MJ, Yamakawa Y et al. Trans activation by the full-length E2 proteins of human papillomavirus type 16 and bovine papillomavirus type 1 in vitro and in vivo: cooperation with activation domains of cellular transcription factors. J Virol 1994; 68: 6655-6666.

31. Etscheid BG, Foster SA, Galloway DA. The E6 protein of human papillomavirus type 16 functions as a transcriptional repressor in a mechanism independent of the tumor suppressor protein p53. Virology 1994; 205:583-585.

32. Akutsu N, Shirasawa H, Asano T et al. p53-dependent and -independent transactivation by the E6 protein of human papillomavirus type 16. J Gen Virol 1996; 77:459-463.

33. Desaintes C, Hallez S, van Alphen P et al. Transcriptional activation of several heterologous promoters by E6 protein of human papillomavirus type 16. J Virol 1992; 66:325-333.

34. Lamberti C, Morrissey LC, Grossman SR et al. Transcriptional activation by the papillomavirus E6 zinc finger protein. EMBO J 1990; 9:1907-1913.

35. Sedman SA, Barbosa MS, Vass WC et al. The full-length E6 protein of human papillomavirus type 16 has transforming and trans-acti-

vating activities and cooperates with E7 to immortalize keratinocytes in culture. J Virol 1991; 65:4860-4866.

36. Carlotti F, Crawford L. Trans-activation of the adenovirus E2 promoter by human papillomavirus type 16 E7 is mediated by retinoblastoma-dependent and -independent pathways. J Gen Virol 1993; 74:2479-2486.

37. Wong HK, Ziff EB. The human papillomavirus type 16 E7 protein complements adenovirus type 5 E1A amino-terminus-dependent transactivation of adenovirus type 5 early genes and increases ATF and Oct-1 DNA binding activity. J Virol 1996; 70:332-340.

38. Massimi P, Pim D, Storey A et al. HPV 16 E7 and adenovirus E1a complex formation with TATA box binding protein is enhanced by casein kinase II phosphorylation. Oncogene 1996; 12:2325-2330

39. Mazzarelli JM, Atkins GB, Geisberg JV et al. The viral oncoproteins Ad5 E1a, HPV16 E7 and SV40 TAg bind a common region of the TBP-associated factor-110. Oncogene 1995; 11:1859-1864.

40. Antinore MJ, Birrer MJ, Patel D et al. The human papillomavirus type 16 E7 gene product interacts with and trans-activates the AP1 family of transcription factors. EMBO J 1996; 15:1950-1960.

41. Gloss B, Chong T, Bernhard HU. Numerous nuclear proteins bind the long control region of human papillomavirus type 16: a subset of 6 of 23 DNase I- protected segments coincide with the location of the cell-type-specific enhancer. J Virol 1989; 63:1142-1152.

42. Nakshatri H, Pater MM, Pater A. Ubiquitous and cell-type specific protein interactions with papillomavirus type 16 and type 18 enhancers. Virology 1990; 178:92-103.

43. Butz K, Hoppe-Seyler F. Transcriptional control of human papillomavirus (HPV) oncogene expression: composition of the HPV type 18 upstream regulatory region. J Virol 1993; 67:6476-6486.

44. Hoppe-Seyler F, Butz K. Cellular control of human papillomavirus oncogene transcription. Mol Carcinogenesis 1994; 10:134-141.

45. Angel P, Karin M. The role of jun, fos and the AP-1 complex in cell-proliferation and transformation. Biochem Biophys Acta 1991; 1072:129-157.

46. Rutberg SE, Saez E, Glick A et al. Differentiation of mouse keratinocytes is accompanied by PKC-dependent changes in AP-1 proteins. Oncogene 1996; 13:167-176.

47. Schenk H, Klein M, Erdbrügger W et al. Distinct effects of thioredoxin and antioxidants on the activation of transcription factor NF-κβ and AP-1. Proc Natl Acad Sci USA 1994; 91:1672-1676.

48. Thierry F, Spyrou G, Yaniv M et al. Two AP-1 sites binding junB are essential for human papillomavirus type 18 transcription in keratinocytes. J Virol 1992; 66:3740-3748.

49. Offord EA, Beard P. A member of the activator protein 1 family found in keratinocytes but not in fibroblasts required for transcription from

a human papillomavirus type 18 promoter. J Virol 1990; 64: 4792-4798.

50. Rösl F, Das BC, Lengert M et al. Antioxidant-induced changes of the AP-1 transcription complex are paralleled by a selective suppression of human papillomavirus transcription. J Virol 1997; 71:362-370.

51. Choo K-B, Huang C-J, Chen C-M et al. Jun-B oncogene aberrations in cervical cancer cell lines. Cancer Letters 1995; 93:249-253.

52. Shrivastava A, Calame K. An analysis of genes regulated by the multi-functional transcriptional regulator Yin Yang-1. Nucleic Acids Res 1994; 22:5151-5155.

53. Bauknecht T, Angel P, Royer H-D et al. Identification of a negative regulatory domain in the human papillomavirus type 18 promoter: interaction with the transcriptional repressor YY1. EMBO J 1992; 11:4607-4617.

54. May M, Dong X-P, Beyer-Finkler E et al. The E6/E7 promoter of extrachromosomal HPV16 DNA in cervical cancers escapes from cellular repression by mutation of target sequences for YY1. EMBO J 1994; 13:1460-1466.

55. Bauknecht T, Jundt F, Herr I et al. A switch region determines the cell type-specific positive or negative action of YY1 on the activity of the human papillomvirus type 18 promoter. J Virol 1995; 69:1-12.

56. Bauknecht T. See RH, Shi Y. A novel C/EBP β-YY1 complex controls the cell-type-specific activity of the human papillomavirus type 18 upstream regulatory region. J Virol 1996; 70:7695-7705.

57. Jundt F, Herr I, Angel P et al. Transcriptional control of human papillomavirus type 18 oncogene expression in different cell lines: role of transcription factor YY1. Virus Genes 1995; 11:53-58.

58. O'Connor MJ, Tan SH, Tan CH et al. YY1 represses human papillomavirus type 16 transcription by quenching AP-1 activity. J Virol 1996; 70:6529-6539.

59. Akira S, Issiki H, Sugita T et al. A nuclear factor for IL-6 expression (NF-IL6) is a member of a C/EBP family. EMBO J 1990; 9:1897-1906.

60. Kyo S, Inoue M, Nishio Y et al. NF-IL6 represses early gene expression of human papillomavirus type 16 through binding to the noncoding region. J Virol 1993; 67:1058-1066.

61. Wang H, Liu K, Yuan F et al. C/EBP-β is a negative regulator of human papillomavirus type 11 in keratinocytes. J Virol 1996; 70: 4839-4844.

62. Klampfer L, Lee TH, Hsu W, Vilcek J, Chen-Kiang S. NF-IL6 and AP-1 cooperatively modulate the activation of the TSG-1 gene by tumor necrosis factor α and interleukin-1. Mol Cell Biol 1994; 14:6561-6569.

63. Hsu W, Kerppola TK, Chen PL et al. Fos and Jun repress transcription activation by NF-IL6 through association at the basic zipper region. Mol Cell Biol 1994; 14:268-276.

64. Gloss B, Bernhard HU. The E6/E7 promoter of human papillomavirus type 16 is activated in the absence of E2 proteins by a sequence-aberrant Sp1 distal element. J Virol 1990; 64:5577-5584.
65. Demeret C, Yaniv M, Thierry F. The E2 transcriptional repressor can compensate for Sp1 activation of the human papillomavirus type 18 early promoter. J Virol 1994; 68:7075-7082.
66. Hoppe-Seyler F, Butz K. Activation of human papilloma virus type 18 E6-E7 oncogene expression by transcription factor Sp1. Nucleic Acids Res 1992; 20:6701-6706.
67. Evans RM. The steroid and thyroid hormone receptor superfamily. Science 1988; 240:889-895.
68. Beato M, Herrlich P, Schütz G. Steroid hormone receptors: many actors in search of a plot. Cell 1995; 83:851-857.
69. Mangelsdorf DJ, Umesono K, Evans RM. The retinoid receptors. In: Sporn MB, Roberts AB, Goodman DS, eds. The Retinoid Receptors: Biology, Chemistry, and Medicine. New York: Raven Press, 1994: 319-349.
70. Chan WK, Klock G, Bernard H-U. Progesterone and glucocorticoid response elements occur in the long control regions of several human papillomaviruses involved in anogenital neoplasia. J Virol 1989; 63:3261-3269.
71. Mittal R, Pater A, Pater MM. Multiple human papillomavirus type 16 glucocorticoid response elements functional for transformation, transient expression, and DNA-protein interactions. J Virol 1993; 67:5656-5659.
72. Medina-Martinez O, Morales-Peza N, Yaniv M et al. A single element mediates glucocorticoid hormone response of HPV18 with no functional interactions with AP1 or hbrm. Virology 1996; 217: 392-396.
73. Pater MM, Mittal R, Pater A. Role of steroid hormones in potentiating transformation of cervical cells by human papillomaviruses. Trends Genet 1994; 2:229-235.
74. Bartsch D, Boye B, Baust C et al. Retinoic acid-mediated repression of human papillomavirus 18 transcription and different ligand regulation of the retinoic acid receptor gene in nontumorigenic and tumorigenic HeLa hybrid cells. EMBO J 1992; 11:2283-2291.
75. Schüle R, Evans RM. Cross-coupling of signal transduction pathways: zinc finger meets leucine zipper. Trends Genet 1991; 7:377-381.
76. Mack DH, Laimins LA. A keratinocyte-specific transcription factor, KRF-1, interacts with AP-1 to activate expression of human papillomavirus type 18 in squamous epithelial cells. Proc Natl Acad Sci USA 1991; 88:9102-9106.
77. Rosenfeld MG. POU-domain transcription factors: pou-er-ful developmental regulators. Genes & Development 1991; 5:897-907.

78. Hoppe-Seyler F, Butz K, zur Hausen H. Repression of the human papillomavirus type 18 enhancer by the cellular transcription factor Oct-1. J Virol 1991; 65:5613-5618.

79. Chong T, Apt D, Gloss B et al. The enhancer of human papillomavirus type 16: binding sites for the ubiquitous transcription factors oct-1, NFA, TEF-2, NF1, and AP-1 participate in epithelial cell-specific transcription. J Virol 1991; 65:5933-5943.

80. O'Connor M, Bernhard H-U. Oct-1 activates the epithelial-specific enhancer of human papillomavirus type 16 via a synergistic interaction with NFI at a conserved composite regulatory element. Virology 1995; 207:77-88.

81. Sibbet GJ, Cuthill S, Campo MS. The enhancer in the long control region of human papillomavirus type 16 is up-regulated by PEF-1 and down-regulated by Oct-1. J Virol 1995; 69:4006-4011.

82. List HJ, Patzel V, Zeidler U et al. Methylation sensitivity of the enhancer from the human papillomavirus type 16. J Biol Chem 1994; 269:11902-11911.

83. Yukawa K, Butz K, Yasui T et al. Regulation of human papillomavirus transcription by the differentiation-dependent epithelial factor Epoc-1/skn-1a. J Virol 1996; 70:10-16.

84. Pederson DS, Heintz NH. Transcription factors and DNA replication. R.G. Landes Company, Austin; 1994.

85. Faus I, Hsu HJ, Fuchs E. Oct-6: a regulator of keratinocyte gene expression in stratified squamous epithelia. Mol Cell Biol 1994; 14:3263-3275.

86. Apt D, Liu Y, Bernhard H-U. Cloning and functional analysis of spliced isoforms of human nuclear factor I-X: interference with transcriptional activation by NFI/CTF in a cell-type specific manner. Nucleic Acids Res 1994; 19:3825-3833.

87. Chong T, Chan W, Bernhard H.-U. Transcriptional activation of human papillomavirus 16 by nuclear factor I, AP1, steroid receptors and a possibly novel transcription factor, PVF: a model for the composition of genital papillomavirus enhancers. Nucleic Acids Res 1990; 18:465-470.

88. Apt D, Chong T, Liu Y et al. Nuclear factor I and epithelial cell-specific transcription of human papillomavirus 16. J Virol 1993; 67:4455-4463.

89. Ishiji T, Lace MJ, Parkkinen S et al. Transcriptional enhancer factor (TEF)-1 and its cell-specific coactivator activate human papillomavirus-16 E6 and E7 oncogene transcription in keratinocytes and cervical carcinoma cells. EMBO J 1992; 11:2271-2281.

90. Dürst M, Kleinheinz A, Hotz M et al. The physical state of human papillomavirus type 16 DNA in benign and malignant genital tumors. J Gen Virol 1985; 66:1515-1522.

91. Choo KB, Pan CC, Han SM. Integration of human papillomavirus type 16 into cellular DNA of cervical carcinoma: preferential deletion of the E2 gene and invariable retention of the long control region and the E6/E7 open reading frames. Virology 1987; 161:259-261.

92. Cullen AP, Reid R, Campion M et al. Analysis of the physical state of different human papillomavirus DNAs in intraepithelial and invasive cervical neoplasm. J Virol 1991; 65:606-612.

93. Schneider-Manoury S, Croissant O, Orth G. Integration of human papillomavirus type 16 DNA sequences: a possible early event in the progression of genital tumors. J Virol 1987; 61:3295-3298.

94. Daniel B, Mukherjee G, Seshadri L et al. Changes in the physical state and expression of human papillomavirus type 16 in the progression of cervical intraepithelial neoplasia lesion analyzed by PCR. J Gen Virol 1995; 76:2589-2593.

95. Schwarz E, Freese K, Gissmann L et al. Structure and transcription of human papillomvirus sequences in cervical carcinoma cells. Nature 1985; 314:111-114.

96. Baker CC, Phelps WC, Lindgren V et al. Structural and transcriptional analysis of human papillomavirus type 16 in cervical carcinoma cell lines. J Virol 1987; 61:962-971.

97. Wilczynski SP, Pearlman L, Walker J. Identification of HPV16 early genes retained in cervical carcinomas. Virology 1988; 166:624-627.

98. Sang B-C, Barbosa MS. Increased E6/E7 transcription in HPV18 immortalized human keratinocytes results from inactivation of E2 and additional cellular events. Virology 1992; 189:448-455.

99. Schwarz A, Schneider-Gaedicke A, zur Hausen H. Human papillomaviruses type-18 transcription in cervical carcinoma cell lines and in human cell hybrids. In: Steinberg BM et al, eds. Papillomaviruses. Cancer Cells 5. Cold Spring Harbor Laboratory, 1987:47-53.

100. El Awady MK, Kaplan JB, O'Brien SJ et al. Molecular analysis of integrated human papillomavirus 16 sequences in the cervical carcinoma cell line SiHa. Virology 1987; 159:389-398.

101. Jeon S, Lambert PF. Integration of human papillomavirus type 16 into the human genome leads to increased stability of E6 and E7 mRNAs: implications for cervical carcinogenesis. Proc Natl Acad Sci USA 1995; 92:1654-1658.

102. Liu Z, Ghai J, Ostrow RS et al. The expression levels of the human papillomavirus type 16 E7 correlate with its transformation potential. Virology 1995; 207:260-270.

103. Elgin SCR. Anatomy of hypersensitive sites. Nature 1984; 309:213-214.

104. Rohdewohld H, Weiher H, Reik W et al. Retrovirus integrations and chromatin structure: Moloney murine leukemia provirus integration sites map near DNAse I-hypersensitive sites. J Virol 1987; 61:336-343.

105. Rösl F, Westphal E-M, zur Hausen H. Chromatin structure and transcriptional regulation of human papillomavirus type 18 DNA in HeLa cells. Mol Carcinogenesis 1989; 2:72-80.

106. Inagaki Y, Tsunokawa Y, Takebe N et al. Nucleotide sequences of cDNAs for human papillomavirus type 18 transcripts in HeLa cells. 1988; J Virol 62:1640-1646.
107. Bauer-Hofmann R, Borghouts C, Auvinen E et al. Genomic cloning and characterization of the nonoccupied allele corresponding to the integration site of human papillomavirus type 16 DNA in the cervical cancer cell line SiHa. Virology 1996; 217:33-41.
108. von Knebel Doeberitz M, Bauknecht T, Bartsch D et al. Influence of chromosomal integration on glucocorticoid-regulated transcription of growth-stimulating papillomavirus genes E6 and E7 in cervical carcinoma cells. Proc Natl Acad Sci USA 1991; 88:1411-1415.
109. Holliday R. The inheritance of epigenetic defects. Science 1987; 238:163-170.
110. Doerfler W. Patterns of DNA methylation—Evolutionary vestiges of foreign DNA inactivation as a host defence mechanism. Biol Chem Hoppe-Seyler 1991; 372:557-564.
111. Antequera F, Macleod D, Bird A. Specific protection of methylated CpGs in mammalian nuclei. Cell 1989; 58:509-517.
112. Rösl F, Arab A, Klevenz B, zur Hausen H. The effect of DNA methylation on gene expression of human papillomaviruses. J Gen Virol 1993; 74:791-801.
113. Thain A, Jenkins O, Clarke AR et al. CpG methylation directly inhibits binding of the human papillomavirus type E2 protein to specific DNA sequences. J Virol 1996; 70:7233-7235.
114. Boyes J, Bird A. DNA methylation inhibits transcription indirectly via a methyl-CpG binding protein. Cell 1991; 64:1123-1134.
115. Levine A, Cantoni G, Razin A. Inhibition of promoter activity by methylation: possible involvement of protein mediators. Proc Natl Acad Sci USA 1991; 88:6515-6518.
116. Meehan RR, Lewis JD, McKay S et al. Identification of a mammalian protein that binds specifically to DNA containing methylated CpGs. Cell 1989; 58:499-507.
117. Antequera F, Boyes J, Bird AP. High levels of de novo methylation and altered chromatin structure at CpG islands in cell lines. Cell 1990; 72:503-514.
118. Razin A, Cedar H. DNA methylation and gene expression. Microbiol Reviews 1991; 55:451-458.
119. Bednarik DP, Cook JA, Pitha PM. Inactivation of the HIV LTR by DNA CpG methylation: evidence for a role in latency. EMBO J 1990; 9:1157-1164.
120. Peng X, Lang CM, Kreider JW. Methylation of cottontail rabbit papillomavirus DNA and tissue-specific expression in transgenic rabbits. Virus Res 1995; 35:101-108.
121. Christy BA, Scangos GA. In vitro methylation of bovine papillomavirus alters its ability to transform mouse cells. Mol Cell Biol 1986; 6:2910-2915.

122. Counts JL, Goodman JI. Alterations in DNA methylation may play a variety of roles in carcinogenesis. Cell 1995; 83:13-15.
123. Butterworth CE. Effect of folate on cervical cancer. Synergism among risk factors. Ann N Y Acad Sci 1992; 669:293-299.
124. Cravo M, Fidalgo P, Pereira AD et al. DNA methylation as an intermediate biomarker in colorectal cancer: modulation by folic acid supplementation. Eur J Cancer Prev 1994; 3:473-479.
125. Dürst M, Petrussevska RT, Boukamp P et al. Molecular and cytogenetic analysis of immortalized human primary keratinocytes obtained after transfection with human papillomavirus 16 DNA. Oncogene 1987; 1:251-256.
126. Pirisi L, Batova A, Jenkins GR et al. Increased sensitivity of human keratinocytes immortalized by human papillomavirus type 16 to growth control by retinoids. Cancer Res 1992; 52:187-193.
127. Kaur P, McDougall JK. HPV18 immortalization of human keratinocytes. Virology 1989; 173:302-310.
128. Dürst M, Seagon S, Wanschura S et al. Malignant progression of an HPV16 immortalized human keratinocyte cell line (HPK 1a) in vitro. Cancer Genet Cytogenet 1995; 85:105-112.
129. Hurlin PJ, Kaur P, Smith PP et al. Progression of human papillomavirus type 18-immortalized human keratinocytes to a malignant phenotype. Proc Natl Acad Sci USA 1991; 88:570-574.
130. DiPaolo JA, Woodworth CD, Popescu NC et al. Induction of human cervical cell carcinoma by sequential transfection with human papillomavirus 16 DNA and viral Harvey ras. Oncogene 1989; 4:395-399.
131. Dürst M, Gallahan D, Jay G et al. Glucocorticoid-enhanced neoplastic transformation of human keratinocytes by human papillomavirus type 16 and an activated ras oncogene. Virology 1989; 173:767-771.
132. Rhim JS, Webber MM, Bello D et al. Stepwise immortalization and transformation of adult human prostate epithelial cells by a combination of HPV18 and v-Ki-ras. Proc Natl Acad Sci USA 1994; 91:11874-11878.
133. de Villiers E-M, Wagner D, Schneider A et al. Human papillomavirus DNA in women without and with cytological abnormalities: results of a five-year follow-up study. Gynecol Oncol 1992; 44:33-39.
134. zur Hausen H. Intracellular surveillance of persisting viral infections: human genital cancer results from deficient cellular control of papillomavirus gene expression. Lancet 1986;ii:489-491.
135. zur Hausen H. Disrupted dichotomous intracellular control of human papillomavirus infection in cancer of the cervix. Lancet 1994; 343:955-957.
136. Stewart N, Bacchetti S. Expression of SV40 large T antigen, but not small t antigen, is required for the induction of chromosomal aberrations in transformed human cells. Virology 1991; 180:49-57.

137. Hashida T, Yasumoto S. Induction of chromosomal abnormalities in mouse and human epidermal keratinocytes by human papillomavirus type 16 E7 oncogene. J Gen Virol 1991; 72:1569-1577.
138. Stanbridge E. Genetic analysis of tumorigenicity in human cell hybrids. Cancer Surveys 1984; 3:335-350.
139. Srivatsan ES, Benedict WF, Stanbridge EJ. Implication of chromosome 11 in the suppression of neoplastic expression in human cell hybrids. Cancer Res 1986; 46:6174-6179.
140. Srivatsan ES, Misra BC, Venugopalan M et al. Loss of heterozygosity for alleles on chromosome 11 in cervical carcinoma. Am J Hum Genet 1991; 49:868-877.
141. Saxon PJ, Srivatsan ES, Stanbridge EJ. Introduction of human chromosome 11 via microcell transfer controls tumorigenic expression of HeLa cells. EMBO J 1986; 5:3461-3466.
142. Koi M, Morita H, Yamada H et al. Normal human chromosome 11 suppresses tumorigenicity of human cervical tumor cell line SiHa. Mol Carcinogenesis 1989; 2:12-21.
143. Smits PHM, Smits HL, Jebbink MF et al. The short arm of chromosome 11 likely is involved in the regulation of the human papillomavirus type 16 early enhancer-promoter and in the suppression of the transforming activity of the viral DNA. Virology 1990; 176:158-165.
144. Smits PHM, Smits HL, Minnaar RP et al. The 55 kDa regulatory subunit of protein phosphatase 2A plays a role in the activation of the HPV16 long control region in human cells with a deletion in the short arm of chromosone 11. EMBO J 1992; 11:4601-4608.
145. Jones PA. DNA methylation errors and cancer. Cancer Res 1996; 56:2463-2467.
146. Rösl F, Dürst M, zur Hausen H. Selective suppression of human papillomavirus transcription in nontumorigenic cells by 5-azacytidine. EMBO J 1988; 7:1321-1328.
147. Little M, Wainwright B. Methylation and p16: Suppressing the suppressor. Nature Med 1995; 1:633-634.
148. Loughran O, Malliri A, Owens D et al. Association of CDKN2A/p16INK4A with human head and neck keratinocyte replicative senescence: relationship of dysfunction to immortality and neoplasia. Oncogene 1996; 13:561-568.
149. von Knebel Doeberitz M, Oltersdorf T, Schwarz E et al. Correlation of modified human papillomavirus early gene expression with altered growth properties in C4-1 cervical carcinoma cells. Cancer Res 1988; 48:3780-3786.
150. von Knebel Doeberitz M, Rittmüller C, zur Hausen H et al. Inhibition of tumorigenicity of cervical cancer cells in nude mice by HPV E6-E7 anti-sense RNA. Int J Cancer 1992; 51:831-834.
151. Crook CP, Morgenstern JP, Crawford L et al. Continued expression of HPV16 E7 protein is required for maintenance of the transformed

phenotype of cells cotransformed by HPV16 plus EJ-ras. EMBO J 1989; 8:513-519.

152. Rösl F, Achtstätter T, Bauknecht T et al. Extinction of the HPV18 upstream regulatory region in cervical carcinoma cells after fusion with nontumorigenic human keratinocytes under nonselective condition. EMBO J 1991; 10:1337-1345.

153. Chen T-M, Pecoraro G, Defendi V. Genetic analysis of in vitro progression of human papillomavirus-transfected human cervical cells. Cancer Res 1993; 53:1167-1171.

154. Seagon S, Dürst M. Genetic analysis of an in vitro model system for human papillomavirus type 16-associated tumorigenesis. Cancer Res 1994; 54:5593-5598.

155. Dürst M, Bosch FX, Glitz D et al. Inverse relationship between HPV16 early gene expression and cell differentiation in nude mouse epithelial cysts and tumors induced by HPV positive human cell lines. J Virol 1991; 65:796-804.

156. Bosch FX, Schwarz E, Boukamp P et al. Suppression in vivo of human papillomavirus type 18 E6-E7 gene expression in nontumorigenic HeLa-fibroblast hybrid cells. J Virol 1990; 64:4743-4754.

157. zur Hausen H. Human papillomaviruses in the pathogenesis of anogenital cancer. Virology 1991; 184:9-13.

158. zur Hausen H, Rösl F. Pathogenesis of cancer of the cervix. Cold Spring Harbor Symposia on Quantitative Biology, 1994 Vol. LIX, pp. 623-628. Cold Spring Harbor Laboratory Press.

159. Khan MA, Jenkins GR, Tolleson, WH et al. Retinoic acid inhibition of human papillomavirus type 16-mediated transformation of human keratinocytes. Cancer Res 1993; 53:905-909.

160. Suzuki T, Okuno H, Yoshida T et al. Difference in transcriptional regulatory function between c-Fos and Fra-2. Nucl Acids Res 1991; 19:5537-5542.

161. Yoshioka K, Deng T, Cavigelli M et al. Antitumor promotion by phenolic antioxidants: inhibition of AP-1 activity through induction of fra expression. Proc Natl Acad Sci USA. 1995; 92:4972-4976.

162. Sinke RJ, Tanigami A, Nakamura Y et al. Reverse mapping of the gene encoding the human fos-related antigen-1 (fra-1) within chromosome band 11q13. Genomics 1993; 18:165.

163. Jesudasan RA, Rahman RA. Chandrashekharappa S et al. Deletion and translocation of chromosome 11q13 sequences in cervical carcinoma cell lines. Am J Hum Genet 1995; 56:705-715.

E6 Protein

Felix Hoppe-Seyler and Martin Scheffner

3.1. INTRODUCTION

HPV DNA is found integrated into the host genome in a high percentage of cervical carcinomas. While integration appears to be random with respect to the host genome, there appears to be some specificity with respect to the viral genome in that integration frequently occurs within the E1 or E2 ORFs (see also chapter 2). As a consequence of this integration the only viral genes that are regularly expressed in cervical cancers are E6 and E7. This observation implies that the E6 and the E7 ORF encode for the major viral oncoproteins. This hypothesis has been corroborated by several lines of evidence. Most importantly, both E6 and E7 have oncogenic properties in various in vitro cell culture systems as well as in transgenic animal models, and inhibition of E6/E7 expression results in growth arrest and reversion of the malignant phenotype of cell lines derived form cervical cancers. To understand why expression of these viral proteins can contribute to, or can induce, immortalization/transformation of infected cells, it is important to identify and characterize both the properties of E6 and E7 involved in immortalization/transformation and the properties that are necessary for the viral life cycle. However, due to the lack of an appropriate cell culture system for HPV propagation,

Papillomaviruses in Human Cancer: The Role of E6 and E7 Oncoproteins, edited by Massimo Tommasino. © 1997 Landes Bioscience.

studies so far have mostly been limited to the characterization of the oncogenic properties of E6 and E7.

As indicated above, the following discussion of the E6 oncoproteins will be confined to those of HPVs associated with anogenital lesions, in particular HPV-16 and HPV-18. The E6 proteins of HPVs associated with cancerous lesions of EV patients will not be addressed due to the limited amount of available experimental data.

3.2. EXPRESSION AND STRUCTURAL ASPECTS

In situ hybridization studies have shown that, in low grade intraepithelial lesions of the cervix, expression of the E6 ORF is restricted to the superficial terminally differentiated layers of the epithelium[1,2] (see also chapter 2). In contrast, in more advanced high grade lesions, as well as in carcinoma in situ and in invasive cancer, the E6 ORF (as well as the E7 ORF) appears to be expressed in all cells. This indicates that, concomitant with the progression of a low grade intraepithelial lesion to a high grade lesion, expression of the viral oncogenes switches from a controlled (low grade lesion) to an uncontrolled or deregulated state (high grade lesions). Thus, it appears that deregulated expression of the viral oncogenes contributes to cellular transformation.

The E6 ORF of HPV-16 and HPV-18 are transcribed from a major promoter (P_{97} and P_{105}, respectively) as a polycistronic message.[3,4] Five different mRNAs containing the E6 ORF have been described but the full-length E6 protein is encoded by only one of these mRNAs. The other four messages encode for truncated versions of E6 as a result of differential splicing either within the E6 ORF or between the E6 ORF and the E1 ORF or the E2 ORF, which are encoded on some of the polycistronic messages (see ref. 5 and references therein). The functional significance of these truncated E6 proteins, however, is not clear since the existence of these putative proteins in HPV-containinig cells has not yet been demonstrated. The most abundant mRNA species found in cervical cancer encodes a truncated E6, commonly termed E6*, and E7.[3,6-8] Because of the presence of the E7 ORF on the spliced E6*-E7 message, it has been speculated that the primary function of the splicing event within the E6 ORF is to facilitate an efficient translation of E7. This hypothesis, however, has recently been challenged by in vitro translation studies suggesting that this is not the case.[9] Thus, the function of the E6*/E7 mRNA still remains to be determined. The unspliced RNA encoding the full-length E6 and E7 is much less abundant, presumably accounting for the low level of E6 protein in HPV containing cells.

The HPV E6 proteins consist of approximately 150 amino acids and have no known intrinsic enzymatic function. They have been detected in the nucleus, in the cytoplasm and membrane fractions.[10,11] E6 proteins are highly conserved among anogenital specific HPVs at the amino acid sequence level. However, similar to their association with clinical lesions, the E6 proteins can be grouped into two classes based on their similarity to each other. "Low risk" E6s show a similarity of approximately 80-90% to each other and, similarly, "high risk" E6s show 70-80% similarity. In contrast, the similarity between the groups is only 50-60%. All E6 proteins contain four CXXC motifs which are presumably involved in zinc binding.[12,13] Since the E7 proteins contain two CXXC motifs with a similar spacing as in the E6 protein, it has been speculated that the E6 and E7 proteins may be evolutionarily related. The overall structure of E6 proteins remains elusive due to the lack of available NMR or X-ray studies.

3.3. IMMORTALIZING AND TRANSFORMING ACTIVITIES

The oncogenic potential of E6 was first demonstrated in immortalization studies using human squamous epithelial cells, the natural host cells of anogenital specific HPVs. Significant with respect to cervical carcinogenesis, these studies revealed that both E6 and E7 are necessary and sufficient to efficiently immortalize these cells.[14-16] Furthermore, it has been recently reported that expression of the E6 oncoprotein in the absence of E7 can interfere with serum and calcium-induced differentiation of human keratinocytes.[17] In keeping with the previous studies, however, continuous proliferation could only be observed upon coexpression of E7. E6 proteins by themselves have been found to immortalize human mammary epithelial cells.[18] In contrast to most other cell culture immortalization/transformation systems, it appears that E6 has a higher oncogenic potential than E7 in this particular system. This may indicate that immortalization of cells is influenced by cell type specific factors and that cells of different origin vary in their susceptibility to different oncogenic stimuli. Finally, E6 can cooperate with an activated *ras* oncogene in the immortalization of baby rodent kidney cells[19,20] and, together with E7, has been shown to induce tumor formation in several transgenic mouse model systems (see below and chapter 4).[21,24]

In correlation with their association with benign lesions, E6 proteins derived from "low risk" HPVs are not, or only weakly, active in immortalization/transformation assays.[25,26] Thus, biochemical and biological characterization of the "high risk" and the "low risk" E6 proteins provides a unique opportunity to correlate biochemical properties of the E6 proteins with their different oncogenic activities.

3.4. MECHANISMS OF IMMORTALIZATION

3.4.1. Interaction with the tumor suppressor protein p53

Mutation of the p53 locus is so far the most common genetic alteration found in a variety of human cancers.[27] The most common pattern seen in cancers is a missense mutation within one allele with the remaining allele being deleted or rearranged. Disruption of both alleles is indicative of tumor suppressor genes, and indeed it has been shown that p53 has cell growth and tumor suppressor properties (for recent reviews, see refs. 28-30).

In its wild-type (wt) conformation, p53 has the activity of a transcriptional modulator. p53 has been shown to bind specifically to DNA and to transactivate genes including p21WAF1, an inhibitor of cyclin-dependent kinases, and bax, which is involved in apoptosis.[31-36] Conversely, p53 expression can also result in repression of certain genes. The mechanism of p53 mediated transcriptional repression has not yet been determined. There is evidence, however, to suggest that this activity of p53 is correlated with its property to bind to the basal transcription factor TBP.[37] Studies of several groups suggest that p53 is activated upon treatment of cells with DNA damaging agents or under hypoxic conditions.[38-41] The activation of p53 is reflected in increased p53 levels and results in either an arrest of the treated cells in the G1/S phase of the cell cycle or apoptosis (Fig. 3.1).[39,42-44] The p53 induced G1 arrest is presumably facilitated by transactivation of the p21WAF1 gene and appears to be important to provide the cell with sufficient time to correct the acquired DNA lesion.[35] Although there is some evidence to suggest that upregulation of bax and downregulation of bcl-2 gene expression by p53 may be involved, the mechanism of p53-induced apoptosis remains to be determined.[45] However, both induction of cell growth arrest and induction of apoptosis would contribute to prevent the accumulation of cells

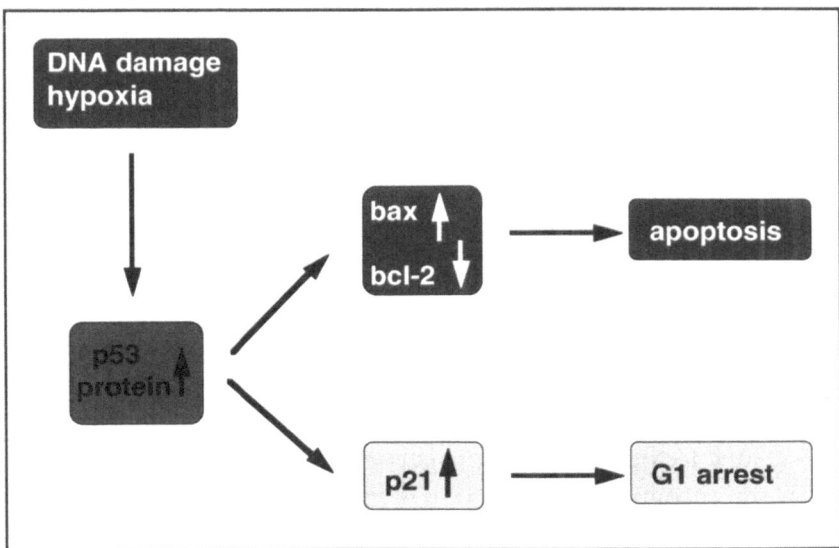

Fig. 3.1. p53-mediated cellular responses to DNA damage and hypoxia. DNA damage and hypoxia result in an increase of intracellular p53 protein levels and induction of apoptosis or cell cycle arrest in G1. The p21WAF1 gene, encoding an inhibitor of cyclin dependent kinases, is transcriptionally activated by p53 and appears to be an important mediator of G1 arrest. Transcriptional activation of bax and transcriptional repression of bcl-2 gene expression by p53 may contribute to induction of apoptosis. Note that p53 may also induce these cellular responses through additional regulatory pathways (see text).

with genomic mutations which may eventually result in cellular transformation.[46] In support of this hypothesis, it has been shown that p53 null cells, or cells expressing mutant p53, have significantly increased mutation rates[47,48] and that p53 null mice are viable but are highly susceptible to development of cancers.[49]

Mutant p53s found in human cancers have lost the properties of wt p53 described above. However, it appears that mutation (or deletion) of the p53 gene is not the only mechanism to inactivate the tumor suppressor properties of p53. In a significant percentage of undifferentiated neuroblastomas and breast cancers, p53 appears to be localized in the cytoplasm rather than in the nucleus.[50,51] The mislocalization presumably results in an inactivation of the wt p53 protein since it cannot exert its function as a transcriptional modulator. Another possible mechanism to inactivate p53 is by interaction with other proteins such as the cellular Mdm2 protein, which has been found overexpressed in certain sarcomas, or viral oncoproteins that negatively interfere with the normal function of p53, as discussed below.[52,53]

3.4.1.1. p53, a common target of DNA tumor viruses

p53 was originally identified as a protein that coimmunoprecipitated with the large tumor antigen (TAg) of simian virus 40 (SV40).[54,55] Subsequently, it was shown that the E1B 55 kDa protein of certain adenovirus (Ad) types is also able to bind to p53.[56] While the significance of these interactions was not clear at the time (p53 was considered to have the properties of a bona fide oncogene), it was suggested that the ability to interact with p53 was related to the oncogenic properties of these viral oncoproteins. With the recognition that p53 has the property of a tumor suppressor protein, it was postulated that the interaction with these viral oncoproteins interferes with the negative regulatory functions of p53. In support of this hypothesis, it could indeed be shown that SV40 TAg and AdE1B inhibit the transcription modulatory functions of p53 and interfere with the biological functions of p53 (i.e., cell cycle arrest, apoptosis) (reviewed in refs. 57-59).

Besides p53, SV40 TAg and E1A, the other oncoprotein of adenovirus, form complexes with another tumor suppressor protein, the product of the retinoblastoma susceptibility gene pRB.[60,61] Since it was shown that the HPV E7 protein also binds to pRB, it was suggested that pRB and p53 might be common targets of small DNA tumor viruses.[62] Accordingly, it was shown that the E6 oncoprotein forms complexes with p53 in vitro.[63] Since the E6 protein of "low risk" HPVs does not, or only weakly, interact with p53, it seems likely that the interaction of the "high risk" E6 proteins with p53 accounts at least in part for their oncogenic potential (see below).

Unlike HPVs, SV40 and adenoviruses have not been etiologically associated with malignant lesions in man, although they can cause malignant tumors in rodents. Several proteins of other viruses with oncogenic potential in man, however, have been reported to interact with p53 (Table 3.1), including the hepatitis B virus X protein and the EBNA5 and BZLF-1 proteins of Epstein-Barr virus.[64-67] The functional significance of these interactions for virus induced cellular transformation, however, is not yet clear.[68,69]

3.4.1.2. E6 induced degradation of p53

Although HPV E6, SV40 TAg, and the Ad E1B 55kDa protein target p53 for inactivation, they do not show any significant similarity at the amino acid sequence level and the consequence of their interaction with p53 seems to be quite different with respect to stability of the p53 protein. In normal cells, wt p53 is a short-lived protein with a half-life ranging from 10-20 minutes (rodent fibroblasts) to 1-2 hours (human primary keratinocytes).[70,71] While the half-life of p53 is greatly extended in SV40 transformed cells, as well as in adenovirus transformed cells, and the steady state level of p53 is accordingly elevated, level and half-life of p53 are decreased in HPV immortalized/transformed cells (in HPV immortalized keratinocytes p53 has a half-life of 15-30 minutes).[71,72] Thus, unlike TAg and the E1B 55 kDa protein which presumably inactivate p53 by sequestering it into complexes, it appears that the E6 oncoprotein inactivates p53 by facilitating its rapid degradation. In support of this hypothesis, it was shown that binding of E6 to p53 induces the degradation of p53 via the ubiquitin-dependent proteolytic pathway (ubiquitin system) in vitro.[73] Since p53 already

Table 3.1. Transforming viruses and their interaction with p53

Virus	Virus family	Factors interacting with p53	Associated human cancers
Human papilloma virus (HPV)	Papovaviridae	E6 (binding through E6-AP)	anogenital cancers, skin cancers and others
Simian virus 40 (SV40)	Papovaviridae	large T (direct binding)	none (tumors in rodents)
Polyomavirus (Py)	Papovaviridae	small t (direct binding)	none (tumors in rodents)
Adenovirus (Ad)	Adenoviridae	E1B 55 kDa (direct binding) E1B 19 kDa (indirect binding) E4orf6 (direct binding)	none (tumors in rodents)
Hepatitis B virus (HBV)	Hepadnaviridae	X (direct binding)	hepatocellular carcinoma
Epstein-Barr virus (EBV)	Herpesviridae	EBNA5 (direct binding) BZLF1 (direct binding)	nasopharyngeal carcinoma, Burkitt's-, Hodgkin's-, B-, and T-cell lymphomas

Compilation of viruses with transforming potential which encode factors interacting with p53.

is a short-lived protein in vivo and since recent evidence suggests that in vivo p53 is a substrate of the ubiquitin system in the absence of E6,[74,75] what is the significance of this observation? As mentioned above, wt p53 is activated upon certain stimuli including genotoxic stress or viral infection (see below). The activation of p53 is generally reflected in increased levels of p53 which is at least in part due to an increased half-life of the protein. In keeping with the property of E6 to induce degradation of p53, it was reported that upon DNA damage levels of p53 do not increase in cells which ectopically express E6.[76,77] This suggests that E6 can circumvent the normal regulation of p53 stability. To fully understand the significance of this observation, however, it will be important to determine the mechanisms by which p53 stability is normally regulated.

The ubiquitin system is a major pathway for selective degradation of cytosolic and nuclear proteins in eukaryotic cells (for recent reviews, see refs. 78 and 79). Over the recent years it has become clear that selective degradation provides a powerful tool to regulate the activity of important cell regulatory proteins. Accordingly, substrates of the ubiquitin system include cyclins, inhibitors of cyclin-dependent kinases, transcription factors including NFkB, *c-jun* and STAT1, the NFkB regulatory protein IkB and, as indicated above, p53. The hallmark of this pathway is the covalent attachment of the 76 amino acid polypeptide ubiquitin to the substrate proteins. Substrate proteins are thereby earmarked for recognition by a ubiquitin-conjugate-specific protease, the so-called proteasome, and subsequently degraded.[80] Ubiquitination of proteins requires the concerted action of at least three classes of protein. These are the ubiquitin-activating enzyme E1, the ubiquitin-conjugating enzymes E2, and the ubiquitin-protein ligases E3. In the first step of ubiquitination, ubiquitin is activated by E1 via the formation of a thioester complex between ubiquitin and E1 (see Fig. 3.2A). The activated ubiquitin is then transferred to one of a number of E2s. Finally, the E2 alone, or in conjunction with an auxiliary factor necessary for substrate recognition, binds to a substrate protein and catalyzes the formation of an isopeptide bond

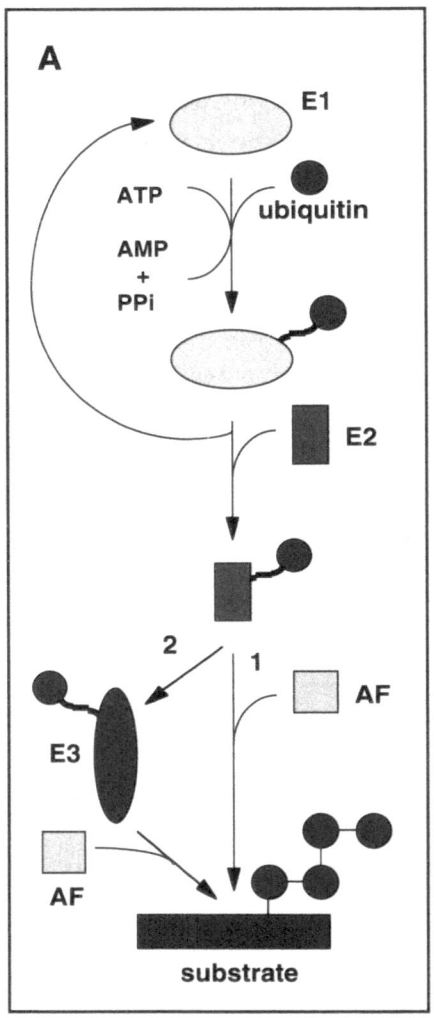

Fig. 3.2. Suggested pathway of ubiquitin conjugation in general and of E6-induced ubiquitination of p53.

(A) Ubiquitin is activated in an ATP-dependent reaction and is covalently linked to ubiquitin activating enzyme E1 via a thioester bond. The activated ubiquitin is then transferred to one of a number of ubiquitin-conjugating enzymes E2. Finally, ubiquitin is attached to a lysine residue of the substrate protein by one of two possibilities. In path 1, ubiquitin is directly transferred from the E2 to the substrate with or without the aid of an auxiliary factor (AF) which facilitates the specific recognition of the substrate by the E2. In path 2, ubiquitin is transferred to one of a number of ubiquitin-protein ligases E3. Ubiquitin is then transferred from the E3 to the substrate. Again, the activity of an AF may be required to facilitate substrate selection. Ubiquitin itself can then serve as a substrate for ubiquitination resulting in multi-ubiquitin chains. Finally, multi-ubiquitinated proteins are recognized by the proteasome and degraded (not shown).

between the C-terminus of ubiquitin and the ε-amino group of a lysine residue of the substrate. Alternatively, the activated ubiquitin is transferred from the E2 to one of a number of E3s.[81,82] In this case, substrate recognition and attachment of ubiquitin to the substrate are presumably mediated by the E3, either alone or in conjunction with an auxiliary factor. The reason for the requirement of E3s in addition to E2s is not clear yet. However, it seems likely that this contributes to ensuring the required selectivity of sub-

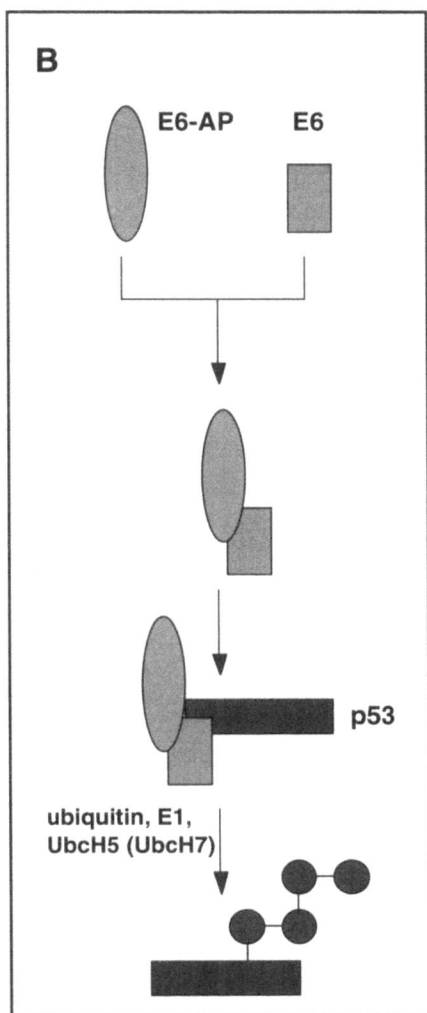

(B) The E6 oncoprotein forms a complex with the ubiquitin-protein ligase E6-AP. The dimeric complex can then bind to p53 which results in E6-AP-catalyzed multi-ubiquitination of p53 in the presence of ubiquitin, E1, and the human E2 UbcH5 (or UbcH7). The lysine residues of p53 that serve as attachment sites for E6-AP-mediated ubiquitination have not been determined, but studies using ubiquitin-aldehyde suggest that ubiquitin is attached to several lysine residues (M. Scheffner, unpublished).

strate recognition and allows the ubiquitin system to finely regulate the turnover of many different proteins (for a detailed discussion, see ref. 83).

The components involved in the E6-induced ubiquitination of p53 were identified in in vitro reconstitution experiments.[84] In addition to ubiquitin, these are E1, an E2 (UbcH5 or UbcH7 in human cells), and the ubiquitin-protein ligase E6-AP (Fig. 3.2B).[85,86] E6-AP was originally identified in a study examining the association of E6 with p53.[87] This study revealed that E6 alone cannot stably interact with p53 but rather requires the function of a 100 kDa cellular protein. This protein can stably interact with E6 in the absence of p53 but not with p53 in the absence of E6 and was, therefore, termed E6-AP for E6-associated protein. In contrast to the "high risk" E6s, and in agreement with the observation that the "low risk" E6 proteins do not or only weakly interact with p53, the "low risk" E6s do not stably interact with E6-AP. A cDNA encoding human E6-AP was cloned but, at the time, the amino acid sequence did not provide insight into the normal function of E6-AP.[88] Biochemical characterization of E6-AP, and its role in E6-dependent ubiquitination of p53, revealed, however, that it has the function of a ubiquitin-protein ligase.[84] Furthermore, E6-AP appears to be a member of a previously undescribed family of ubiquitin-protein ligases. Members of this family are characterized by a C-terminal region of approximately 350 amino acids in length termed the hect domain (*homologous to E6-AP C terminus*).[81,82] The hect domain probably represents the catalytic domain of these otherwise unrelated proteins since recent studies have demonstrated that the hect domain is necessary and sufficient to form thioester complexes with ubiquitin (see Fig. 3.2) (S.E. Schwarz and M. Scheffner, unpublished).

3.4.1.3. Functional consequences of the E6-p53 interaction

The in vitro characterization of the interaction of E6 with p53 was complemented by studies investigating the influence of E6 expression on intracellular p53 activities. As mentioned above, HPV-positive cancer cells generally contain low but detectable levels of p53 protein which is most likely due to E6-induced degradation of p53.[71,89] Beside the effect on p53 stability, it was shown in transient transfection assays that HPV16 and HPV18 E6 inhibit both the transcriptional activator and the transcriptional repressor function of wt p53, thus interfering with functions which are considered to be important for p53-mediated tumor suppression.[90-95] It is noteworthy that the interference of E6 with p53-mediated transcriptional regulation does not necessarily require degradation of p53, as the DNA binding and transactivation functions of p53 were also inhibited by mutant HPV18 E6 proteins defective for induction of p53 degradation.[96,97] Furthermore, the transcriptional modulatory activities of p53 were only weakly inhibited by the E6 proteins of "low risk" HPV types. These results further correlate the ability of individual E6 proteins to interact with p53 and their different transforming potential.

Expression of HPV16 E6 has been shown to result in increased mutagenesis and genetic instability in different experimental cell systems, such as in normal human fibroblasts, human uroepithelial cells, or rat colon carcinoma cells with intact p53.[98-100] This is consistent with the idea that "high risk" E6 proteins contribute to cell transformation by interfering with the function of p53 as stabilizer of the genome. Interestingly, "low risk" HPV11 E6 also induced a slight elevation in mutation rate in one study,[100] which may be related to its ability to weakly bind p53 and to interfere with the DNA binding and transactivation function of p53 to a low degree.[93,101] The mutagenic effects of "high risk" E6 proteins in these systems are likely to be linked to the inhibition of the normal cellular response to DNA damage. In line with this notion, expression of "high risk" E6 can inhibit several features of an intact cellular response to genotoxic stress. In experimental cell systems, where "high risk" E6 was ectopically expressed from heterologous promoters, no increase in intracellular p53 protein levels or induction of cell cycle arrest in G1 following genotoxic stress could be observed (see above).[76,77] Furthermore, it is of considerable interest that the genetic instability induced by "high risk" E6 may also contribute to the integration of "high risk" HPV DNA into the host chromosome.[102]

Taken together, these results suggest that expression of a "high risk" E6 protein may have the same functional consequences as inactivation of the p53 gene by somatic mutation, namely loss of p53-mediated transcriptional regulation, as observed for most p53 mutants, and inhibition of the normal cellular response to DNA damage. In line with this notion was the finding that a series of cervical cancer cell lines either contained HPV sequences or mutations within the p53 gene.[72,103] This issue was complicated, however, by studies analyzing the correlation between the p53 and HPV status in primary tumors. Such studies indicated that p53 mutations are rare events in both HPV-positive and HPV-negative cancers.[104-106] However, the possibility that some of the cancers scoring HPV-negative by available detection systems may contain thus far unidentified HPV types, has to be considered.

The simple view that HPV-positive cancer cells lack functional p53 protein and do not exhibit cellular responses to genotoxic stress was further complicated by the observation that several HPV-positive cancer cell lines contain endogenous p53 protein competent for transcriptional activation and exhibit features of an intact cellular response to DNA damaging agents.[107] A possible explanation for the differences between the situation in experimental cell systems and HPV-positive cancer cells could be provided by the different expression levels of E6, which are believed to be low in HPV-positive cancer cells. Thus, the levels of E6 protein in established cervical cancer cells may limit its capacity to interfere with p53 function. In addition, it was recently shown that E6/E7 expression in HPV-positive cancer cells is repressed by DNA damaging agents such as cisplatin and mitomycin C.[108] A reduction in E6 expression level may further increase the functional p53 pool in these cells and augment p53-associated cellular responses to genotoxic anticancer agents. These findings also indicate that regulatory pathways downstream of p53 are principally intact in HPV-positive cancer cells and can be induced by certain stimuli. Consistent with this notion, it has been shown that expression of the bovine papillomavirus type 1 (BPV1) E2 protein, an inhibitor of HPV16 and HPV18 E6/E7 transcription, led to the activation of p53-associated growth inhibitory pathways in HPV18-positive HeLa cells.[109,110] In addition, overexpression of p53 in HPV-positive cervical cancer cells led to apoptosis indicating that apoptotic pathways downstream of p53 are also intact.[111,112] This is further supported by the observation that treatment of HPV-positive cancer cell lines with genotoxic agents ultimately resulted in apoptotic cell death.[108] It is not yet known if, and to which extent, p53-independent pathways may also contribute to the response of HPV-positive cancer cells to DNA damaging agents since induction of both growth arrest genes and apoptosis can also be p53-independent.[113-115] Nonetheless, the observation that apoptosis is inducible in HPV-positive cancer cells by genotoxic anticancer agents, may have clinical implications since it could provide a molecular explanation for the therapeutic effects of irradiation or chemotherapeutic agents, such as cisplatin, in the treatment of cervical cancer.

3.4.2. p53-independent properties of E6

In several experimental systems, HPV E6 proteins exhibit transforming and growth deregulating activities which are independent of the interference with p53. For example, mutant HPV16 and 18 E6 proteins defective for interacting with p53 were able to cooperate with EJ-*ras* to immortalize baby mouse kidney cells.[96] In addition, the transforming activity of "high risk" HPV16 E6, in human mammary epithelial and mouse NIH3T3 cells, appears to be linked to additional targets besides p53 as transdominant p53 mutants could not substitute for E6 in transformation.[26,71] Furthermore, a mutant HPV16 E6 protein, which does not detectably interact with p53, promoted the proliferation of HPV18-positive SW756 cells, in which the endogenous E6/E7 expression was conditionally suppressed.[116] Similarly, the growth promoting activities of HPV16 E6 in a conditionally immortalized mouse cell model were reported to be independent of interfering with p53 activities.[117] In this context, it is also noteworthy that HPV types linked to skin cancers associated with the rare hereditary disease, epidermodysplasia verruciformis, such as HPV5 and HPV8, express E6 proteins which do not detectably interact with p53, although they exhibit transforming activities in experimental systems.[118-120]

Transforming activities of different E6 proteins may also be influenced by the cellular environment (see also above). For example, the E6 proteins of HPV6 and BPV1 exhibited transforming activities in human mammary epithelial cells.[26] Interestingly, the half-life of p53 in the immortalized cells was markedly decreased which appears to be contradictory to the finding that both BPV1 and HPV6 E6 do not detectably induce p53 degradation in vitro. This contradiction may suggest, however, that only a particular cellular environment may support p53 degradation by these E6 proteins or that the p53 half-life in these cells is reduced by E6-independent pathways. Cell type specificity of E6 function was also observed in another study, where "high risk" HPV16 E6 was competent to bypass the so called M1 stage of cellular senescence in human mammary epithelial cells but not in fibroblasts.[121]

Several functional aspects of the "high risk" E6 protein may contribute to its p53-independent transforming activity. It has been reported that HPV16 E6 can both transcriptionally activate, and repress a series of heterologous promoters, and the latter activity has been shown to be p53-independent.[122-124] Thus, it is possible that "high risk" E6 can contribute to cellular transformation by deregulating transcription of unidentified growth regulatory genes. However, "low risk" HPV6 E6 protein has also been shown to transactivate an adenovirus E2 test promoter.[89] This suggests that either the transactivating function is not required, or sufficient, for the transforming function of E6 or that "high risk" E6 may in addition regulate a different set of target promoters than "low risk" E6.

Experiments employing a transgenic mouse model to analyze the influence of HPV16 E6 and E7 genes on mouse lens development indicated that E6 also can reduce apoptosis through p53-independent mechanisms.[22,125] This regulation may also play a role in the inhibition of the E7-induced apoptosis, and thus may counteract the apoptotic elimination of cells with growth deregulation (see below).

HPV16 E6, and to a lower degree HPV6 E6, has recently been shown to activate telomerase, an enzymatic RNA-protein complex counteracting the progressive loss of chromosomal telomeres during normal cellular senescence.[126,127] Interestingly, induction of telomerase by HPV16 E6 was independent of its ability to target p53 for degradation and was cell type specific, being detectable in early passage human keratinocytes and mammary epithelial cells

but not in fibroblasts.[126] Telomerase activity in HPV16 E6 expressing cells was associated with an extended lifespan but not with
immortalization. These results indicate that telomerase activation
by E6 in epithelial cells may play an early role in HPV-associated
cell transformation. This interpretation, however, is somewhat
complicated by recent studies showing that primary and passaged
keratinocytes from normal skin and cervical epithelium exhibited
telomerase activities per se, most likely due to enzymatic activity
in proliferating basal cells, the putative target cell for HPV
infection.[128,129]

Finally, it is noteworthy that the E6 protein of "high risk" HPV
types may contribute to cellular transformation by associating with
cellular proteins other than E6-AP/p53. Using the yeast two hybrid system, an E6 binding protein (E6BP) was isolated from a
HeLa cell expression library, which specifically interacted with "high
risk" but not with "low risk" HPV E6 proteins.[130] E6BP is identical
to the putative calcium binding protein ERC-55. Although the biological effects of the interaction between E6 and E6BP have yet to
be determined, they may involve the deregulation of calcium-as-
sociated pathways involved in differentiation, cell cycle control, or
apoptosis of HPV-infected keratinocytes. Additional proteins, including an unknown protein kinase and its substrate (pp182), have
also been demonstrated to associate with HPV E6 proteins.[131] Further work is required to identify and characterize these factors in
detail.

3.4.3. Functional Interplay Between the E6 and E7 Oncoproteins

Although both the E6 and E7 oncoproteins exhibit transforming activities per se, their oncogenic potential is strongly increased
when both factors are coexpressed.[14-16] This indicates that the viral
factors may functionally cooperate in the process of cell transformation. A potential model for this cooperation was initially
deduced from experiments analyzing the cooperation between
the adenovirus transforming protein E1A and E1B. Similar to the
"high risk" HPV E6 and E7 proteins, E1B and E1A target p53 and
pRb for functional inactivation, respectively.[132] It has been shown

that cellular growth deregulation through E1A alone results in an increase of p53 protein levels and induction of apoptosis in cells devoid of E1B.[133,134] Thus, cells can respond to growth deregulatory signals by inducing p53-mediated cellular suicide, resulting in the elimination of cells with uncontrolled growth behavior. In contrast, when the E1B protein was coexpressed with E1A, apoptosis was inhibited, most likely through the interference of E1B with p53. This results in the outgrowth of cells with deregulated proliferative behavior. The idea that, similar to the interplay between E1B and E1A, "high risk" E6 can potentiate the growth deregulatory effect of E7 by inhibiting p53-mediated apoptosis, is supported experimentally. Expression of the HPV16 E7 gene in a transgenic mouse model resulted in increased apoptosis, which was counteracted by HPV16 E6 (see also chapter 4).[22,125] Interestingly, a p53 null genotype acted additively with the E6 transgene, indicating that E6 can inhibit E7-induced apoptosis both by p53-dependent and p53-independent pathways.[125] The results of this mouse model were validated and extended by studies in human uroepithelial cells, showing that, following DNA damage, E7 primes these cells for apoptosis which is counteracted by E6.[135] Thus, the functional cooperation between "high risk" E6 and E7 proteins in the process of HPV-associated cell transformation may be, at least partly, due to the E6-mediated inhibition of the cellular suicide program which otherwise eliminates cells exposed to deregulatory growth signals, including those exerted by "high risk" E7 (Fig. 3.3).

E6 and E7 functions also appear to be linked at the level of deregulation of cell cycle and growth control. In this scheme, pRb is a common target for both viral oncoproteins, being inhibited by physically associating with "high risk" E7 protein (see chapter 4) and by E6-induced alterations of its phosphorylation status. In this scenario, "high risk" E6 inhibits p53-mediated transcriptional activation of the cyclin-dependent kinase (CDK) inhibitor p21WAF1.[33] p53 induction of p21WAF1 appears to contribute to the maintenance of the underphosphorylated status of pRb, which is required for the function of pRb as an inhibitor of unscheduled cell cycle progression through the G1 phase.[136] The inhibition of

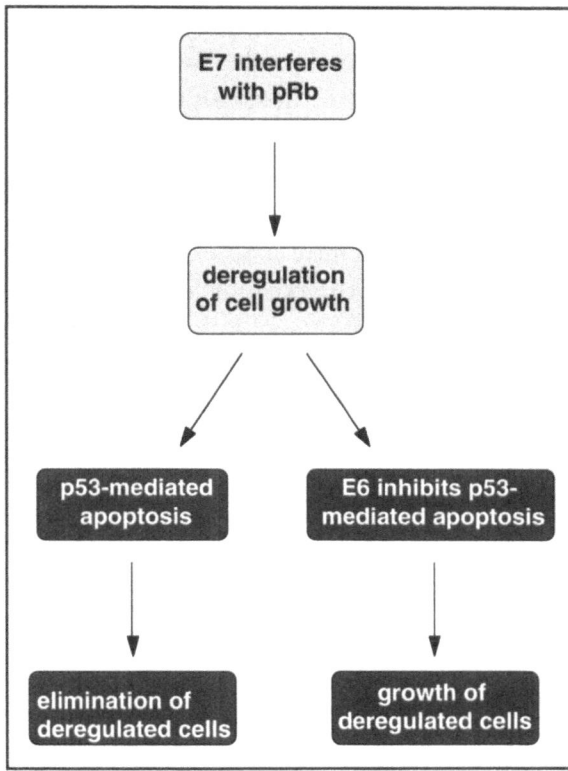

Fig. 3.3. Cooperation between HPV E6 and E7 oncoproteins during cellular transformation. Growth deregulation by E7 can be counteracted by p53-mediated apoptosis, resulting in the elimination of cells with altered growth behavior. By interfering with p53-mediated apoptosis, E6 inhibits the elimination of cells with disturbed growth control. This results in outgrowth of deregulated cells which, in addition, are genetically unstable through E6-mediated inhibition of p53-mediated genomic stabilization.

p53-mediated induction of p21WAF1 by "high risk" E6 may thus result in a disturbance of pRb control and act in concert with pRb inactivation through E7 binding.

3.5. ROLE OF E6 IN THE VIRAL LIFE CYCLE

The life cycle of human papillomaviruses is closely linked to the differentiation status of infected keratinocytes. E6/E7 gene expression is barely detectable in the lower, undifferentiated layers of the epithelium and increases concomitantly with keratinocyte differentiation in the upper layers. Vegetative DNA replication of human papillomaviruses occurs only in the differentiated upper layers of the epithelium, in which the keratinocytes no longer cycle and divide. Since HPVs are dependent on host cell factors for the replication of their DNA, they have presumably evolved strategies to activate cellular pathways necessary for DNA replication in nonreplicating cells. By abrogating the cell cycle checkpoint pRb, E6 and E7 could allow the recruitment of host cell factors necessary for viral DNA replication. In addition, E6 may also counteract the elevation of cellular p53 concentrations in response to the effects of E7, thus blocking p53-mediated apoptosis of virally infected cells.[137] Finally, p53 might also have a role in prohibiting unscheduled DNA replication outside the normal cell cycle which could be inhibited by E6 and, thus, allowing viral replication.[138] All these activities may contribute to the ability of HPVs to induce viral DNA replication in otherwise quiescent cells. Clearly, however, the situation is likely to be more complex. HPVs have presumably evolved additional functions to allow their propagation independently of interfering with p53 and pRb, as indicated by the fact that "low risk" HPVs can efficiently replicate in terminally differentiated keratinocytes, yet their E6 and E7 proteins only poorly interfere with p53 and pRb.

3.6. THERAPEUTIC PERSPECTIVES

The fact that both E6 and E7 are regularly expressed in HPV-positive cancer cells makes them an attractive target for the treatment of HPV-associated cancers (for immunological aspects see chapter 5). Indeed, it has been shown that continuous E6/E7 expression is necessary for the maintenance of the transformed phenotype of cervical cancer cell lines in vitro and in vivo.[139,140] Possible therapeutic interventions to block E6/E7 functions can be envisioned on several biological levels (Fig. 3.4). One could attempt to specifically block HPV E6/E7 transcription. For example, due to the mode of viral integration into the host chromosome, the HPV E2 protein often is absent in cancers. E2 has been shown to inhibit the activity of the HPV16 or HPV18 E6/E7 promoter by binding to specific DNA recognition sequences within the viral transcriptional control region.[141,142] Thus, E2 may provide a tool to specifically inhibit E6/E7 expression in HPV-positive cancers, although recent evidence indicates that repression of E6/E7 transcription by E2 may not be efficient for chromosomally integrated viral sequences.[109,143]

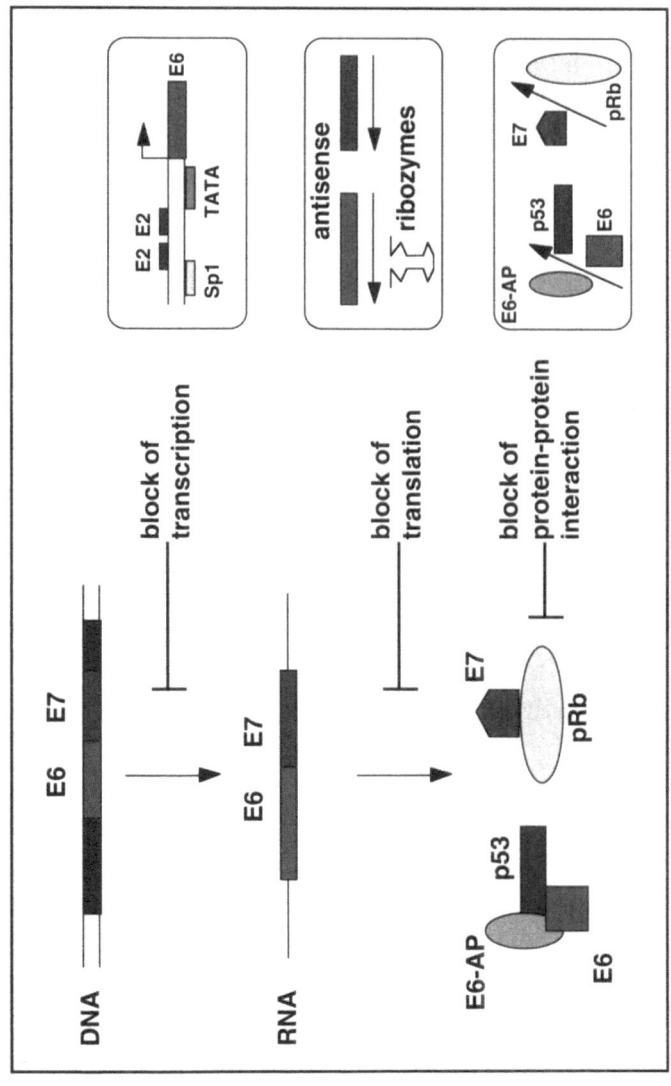

Fig. 3.4. Possible levels of therapeutic intervention with HPV E6/E7 oncogene expression. E6/E7 expression may be blocked at the transcriptional level, e.g., by the transcriptional repressor E2, which binds close to the TATA box of the E6/E7 promoter and sterically interferes with the formation of a transcriptional initiation complex. E6/E7 gene activity could be blocked at the RNA level, e.g., by antisense or ribozyme constructs targeting E6 and/or E7 RNA processing. E6 and E7 function also could be blocked at the protein level by molecules interfering with the interaction between E6 and E6AP or E7 and pRb, respectively.

Antisense sequences directed against E6/E7 transcripts can inhibit E6/E7 expression at the RNA level and, thus, could possess potential for future approaches to gene therapy of HPV-associated cancers.[139,144-150] It should be noted that it is not yet clear if the observed antitumorigenic effects of antisense oligonucleotides is attributable to either E6 or E7 inhibition, as both E6 and E7 are translated from a common mRNA.

Given that the interference of "high risk" E6 with p53 is necessary for the maintenance of the transformed phenotype, and does not represent a transient early event, one could envisage several targets for therapeutic intervention of E6 function on the protein level. For example, one could attempt to block the interaction between E6 and the cellular E6-AP protein which is required for the degradation of p53. Indeed, a short peptide derived from E6-AP has been identified which can block the binding of E6 to E6-AP.[151] Such inhibition of the interaction between E6 and E6-AP could increase the functional p53 pool in HPV-positive cancer cells which, in turn, may result in p53-associated growth arrest or in induction of apoptosis. Furthermore, augmenting p53 function in HPV-positive cancer cells by blocking E6 may also have therapeutic implications for combinatorial therapy regimens. In this scenario, increased p53 function may support the cytotoxic effects of genotoxic agents already used in the treatment of cervical cancer, such as radiation or chemotherapy, by more efficiently inducing apoptotic cell death in response to the DNA damage introduced by these substances. In addition to these approaches, it also will be an important future task to elucidate the p53-independent transforming functions of E6, which may also represent potential targets for specific inhibition of HPV-associated oncogenicity.

REFERENCES

1. Dürst M, Glitz D, Schneider A et al. Human papillomavirus type 16 (HPV16) gene expression and DNA replication in cervical neoplasia: analysis by in situ hybridization. Virology 1992; 189:132-140.
2. Stoler MH, Rhodes, CR, Whitbeck SM et al. Human papillomavirus type 16 and 18 gene expression in cervical neoplasias. Hum Pathol 1993; 23:117-128.

3. Smotkin D, Wettstein FO. Transcription of human papillomavirus type 16 early genes in cervical cancer and a cervical cancer derived cell line and identification of the E7 protein. Proc Natl Acad Sci USA 1986; 83:4680-4684.

4. Thierry F, Yaniv, M. The BPV-1 E2 trans-acting protein can be either an activator or repressor of the HPV-18 regulatory region. EMBO J 1987; 6:3391-3397.

5. Shally M, Alloul N, Jackman A et al. The E6 variant proteins E6I-E6IV of human papillomavirus 16: expression in cell free systems and bacteria and study of their interaction with p53. Virus Res 1996; 42:81-96.

6. Schwarz E, Freese UK, Gissmann L et al. Structure and transcription of human papillomavirus sequences in cerviacla carcinoma cells. Nature 1985; 314:111-114.

7. Baker CC, Phelps WC, Lindgren V et al. Structural and translational analysis of human papillomavirus type 16 sequences in cervical carcinoma cell lines. J Virol 1987; 61:962-971.

8. Smotkin D, Prokoph H, Wettstein FO. Oncogenic and nononcogenic human genital papillomaviruses generate the E7 mRNA by different mechanisms. J Virol 1989; 63:1441-1447.

9. Stacey SN, Jordan D, Snijders PJ et al. Translation of the human papillomavirus type 16 E7 oncoprotein from bicistronic mRNA is independent of splicing events within the E6 open reading frame. J Virol 1995; 69:7023-7031.

10. Grossman SR, Mora R, Laimins LA. Intracellular localization and DNA-binding properties of human papillomavirus type 18 E6 protein expressed with a baculovirus vector. J Virol 1989; 63:366-374.

11. Liang XH, Volkmann M, Klein R et al. Co-localization of the tumor-suppressor protein p53 and human papillomavirus E6 protein in human cervical carcinoma cell lines. Oncogene 1993; 8:2645-2652.

12. Barbosa MS, Lowy DR, Schiller JT. Papillomavirus polypeptides E6 and E7 are zinc-binding proteins. J Virol 1989; 63:1404-1407.

13. Grossman SR, Laimins LA. E6 potein of human papillomavirus type 18 binds zinc. Oncogene 1989; 4:1089-1093.

14. Hawley-Nelson P, Vousden KH, Hubbert NL et al. HPV 16 E6 and E7 proteins cooperate to immortalize human foreskin keratinocytes. EMBO J 1989; 8:3905-3910.

15. Münger K, Phelps WC, Bubb V et al. The E6 and E7 genes of the human papilloamvirus type 16 are necessary and sufficient for transformation of primary human keratinocytes. J Virol 1989; 63:4417-4421.

16. Watanabe S, Kanda T, Yoshiike K. Human papillomavirus type 16 transformation of primary human embryonic fibroblasts requires expression of open reading frames E6 and E7. J Virol 1989; 63:965-969.

17. Stöppler MC, Ching K, Stöppler H et al. Natural variants of the human papillomavirus type 16 E6 protein differ in their abilities to alter keratinocyte differentiation and to induce p53 degradation. J Virol 1996; 70:6987-6993.

18. Band V, DeCaprio JA, Delmolino L et al. Loss of p53 protein in human papillomavirus type 16 E6-immortalized human mammary epithelial cells. J Virol 1991; 65:6671-6676.

19. Storey A, Banks L. Human papillomavirus type 16 E6 gene cooperates with EJ-ras to immortalize primary mouse cells. Oncogene 1993; 8:919-924.

20. Liu Z, Ghai J, Ostrow RS et al. The E6 gene of human papillomavirus type 16 is sufficient for transformation of baby rat kidney cells in cotransfection with activated Ha-ras. Virology 1994; 201:388-396.

21. Lambert PF, Pan H, Pitot HC et al. Epidermal cancer associated with expression of human papillomavirus type 16 E6 and E7 oncogenes in the skin of transgenic mice. Proc Natl Acad Sci USA 1993; 90:5583-5587.

22. Pan H, Griep AE. Altered cell cycle regulation in the lens of HPV-16 E6 or E7 transgenic mice: implications for tumor suppressor gene function in development. Genes Dev 1994; 8:1285-1299.

23. Arbeit JM, Münger K, Howley PM et al. Progressive squamous epitelial neoplasia in K14-human papillomavirus type 16 transgenic mice. J Virol 1994; 68:4358-4368.

24. Arbeit JM, Howley PM, Hanahan D. Chronic estrogen-induced cervical and vaginal squamous carcinogenesis in human papillomavirus type 16 transgenic mice. Proc Natl Acad Sci USA 1996; 93:2930-2935.

25. Halbert CL, Demers GW, Galloway DA. The E6 and E7 genes of human papillomavirus type 6 have weak immortalizing activity in human epithelial cells. J Virol 1992; 66:2125-2134.

26. Band V, Dalal S, Delmolino L et al. Enhanced degradation of p53 protein in HPV-6 and BPV-1 immortalized human mammary epithelial cells. EMBO J 1993; 5:1847-1852.

27. Greenblatt MS, Bennett WP, Hollstein M et al. Mutations in the p53 tumor suppressor gene: clues to cancer etiology and molecular pathogenesis. Cancer Res 1994; 54:4855-4878.

28. Prives C. How loops, β sheets, and α helices help us to understand p53. Cell 1994; 78:543-546.

29. Deppert W. The yin and yang of p53 in cellular proliferation. Semin Cancer Biol 1994; 5:187-202.

30. Haffner R, Oren M. Biochemical properties and biological effects of p53. Curr Opin Genet Dev 1995; 5:84-90.

31. Kern SE, Pietenpol JA, Thiagalingam S et al. Oncogenic forms of p53 inhibit p53-regulated gene expression. Science 1992; 256:827-830.

32. Funk WD, Pak Dt, Karas RH et al. A transcriptionally active DNA-binding site for human p53 protein complexes. Mol Cell Biol 1992; 12:2866-2871.

33. El-Deiry WS, Tokino T, Velculescu VE et al. WAF-1, a potential mediator or p53 tumor suppression. Cell 1993; 75:817-825.
34. Harper JW, Adami GR, Wei N et al. The p21 CDK-interacting protein cip1 is a potent inhibitor of G1 cyclin-dependent kinases. Cell 1993; 75:805-816.
35. Deng C, Zhang P, Harper JW et al. Mice lacking p21CIP1/WAF1 undergo normal development, but are defective in G1 checkpoint control. Cell 1995; 82:675-684
36. Miyashita T, Reed JC. Tumor suppressor p53 is a direct transcriptional activator of the human bax gene. Cell 1995; 80:293-299.
37. Horikoshi N, Usheva A, Chen J et al. Two domains of p53 interact with the TATA-binding protein, and the adenovirus 13S E1A protein disrupts the association, relieving p53-mediated transcriptional repression. Mol Cell Biol 1995; 15:227-234.
38. Maltzman W, Czyzyk L. UV irradiation stimulates levels of p53 cellular tumor antigen in nontransformed mouse cells. Mol Cell Biol 1984; 4:1689-1694.
39. Kastan MB, Zhan Q, El-Deiry WS et al. A ammalian cell cycle checkpoint pathway utilizing p53 and GADD45 is defective in Ataxia-telangiectasia. Cell 1992; 71:587-597.
40. Fritsche M, Haessler C, Brandner G. Induction of nuclear accumulation of the tumor-suppressor protein p53 by DNA-damaging agents. Oncogene 1993; 8:307-318.
41. Graeber TG, Osmanian C, Jacks T et al. Hypoxia-mediated selection of cells with diminished apoptotic potential in solid tumors. Nature 1996; 379:88-91.
42. Lowe SW, Schmitt EM, Smith SW et al. p53 is required for radiation-induced apoptosis in mouse thymocytes. Nature 1993; 362: 847-849.
43. Caelles C, Helmberg A, Karin M. p53-dependent apoptosis in the absence of transcriptional activation of p53-target genes. Nature 1994; 370:220-223.
44. Canman CE, Gilmer TM, Coutts SB et al. Growth factor modulation of p53-mediated growth arrest versus apoptosis. Genes Dev 1995; 9:600-611.
45. Miyashita T, Krajewski S, Krajewska M et al. Tumor suppressor p53 is a regulator of bcl-2 and bax gene expression in vitro and in vivo. Oncogene 1994; 9:1799-1805.
46. Lane DP. p53, guardian of the genome. Nature 1992; 358:15-16.
47. Livingstone LR, White A, Sprouse J et al. Altered cell cycle arrest and gene amplification potential accompany loss of wild-type p53. Cell 1992; 70:923-935.
48. Yin Y, Tainsky MA, Bischoff FZ et al. Wild-type p53 restores cell cycle control and inhibits gene amplification in cells with mutant p53 alleles. Cell 1992; 70:937-948.

49. Donehower LA, Harvey M, Slagle BL et al. Mice deficient for p53 are developmentally normal but susceptible to spontaneous tumors. Nature 1992; 356:215-221.

50. Moll UM, Riou G, Levine AJ. Two distinct mechanisms alter p53 in breast cancer: Mutation and nuclear exclusion. Proc Natl Acad Sci USA 1992; 89:7262-7266.

51. Moll UM, LaQuaglia M, Bernard J et al. Wild-type p53 protein undergoes cytoplasmic sequestration in undifferentiated neuroblastomas but not in differentiated tumors. Proc Natl Acad Sci USA 1995; 92:4407-4411.

52. Momand J, Zambetti GP, Olson DC et al. The mdm-2 oncogene product forms a complex with the p53 protein and inhibits p53-mediated transactivation. Cell 1992; 69:1237-1245.

53. Oliner JD, Kinzler KW, Meltzer PS et al. Amplification of a gene encoding a p53-associated protein in human sarcomas. Nature 1992; 358:80-83.

54. Lane DP, Crawford LV. T antigen is bound to a host protein in SV40-transformed cells. Nature 1979; 278:261-263.

55. Linzer DIH, Levine AJ. Characterization of a 54K dalton cellular SV40 tumor antigen present in SV40-transformed cells and uninfected embryonal carcinoma cells. Cell 1979; 17:43-52.

56. Sarnow P, Ho YS, Williams J et al. Adenovirus E1b-58kd tumor antigen and SV40 large tumor antigen are physically associated with the same 54kd cellular protein in transformed cells. Cell 1982; 28:387-394.

57. Yew PR, Berk AJ. Inhibition of p53 transactivation required for transformation by adenovirus early 1B protein. Nature 1992; 357:82-85.

58. Ludlow JW. Interactions between SV40 large tumor antigen and the growth suppressor proteins pRb and p53. FASEB J 1993; 7:866-871.

59. White E. Regulation of p53-dependent apoptosis by E1A and E1B. Curr Top Microbiol Immunol 1995; 199:34-58.

60. Whyte P, Buchkovich KJ, Horowitz JM et al. Association between an oncogene and an antioncogene: the adenovirus E1a proteins bind to the retinoblastoma gene product. Nature 1988; 334:124-129.

61. DeCaprio JA, Ludlow JW, Figge J et al. SV40 alrge tumor antigen forms a specific complex with the product of the retinoblastoma susceptibility gene. Cell 1988; 54:275-283.

62. Dyson N, Howley PM, Münger K et al. The human papillomavirus-16 E7 oncoprotein is able to bind to the retinoblastoma gene product. Science 1989; 243:934-937.

63. Werness BA, Levine AJ, Howley PM. Association of human papillomavirus types 16 and 18 E6 proteins with p53. Science 1990; 248:76-79.

64. Feitelson MA, Zhu M, Duan X-L et al. Hepatitis X-antigen and p53 are associated in vitro and in liver tissues from patients with primary hepatocellular carcinoma. Oncogene 1993; 8:1109-1117.

65. Wang XW, Forrester K, Yeh H et al. Hepatitis B virus X protein inhibits p53 sequence-specific DNA binding, transcriptional activity, and association with transcription factor ERCC3. Proc Natl Acad Sci USA 1994; 91:2230-2234.

66. Zhang Q, Gutsch D, Kenney S. Functional and physical interactions between p53 and BZLF-1: implications for Epstein-Barr virus latency. Mol Cell Biol 1994; 14:1929-1938.

67. Szekely L, Selivanova G, Magnusson KP et al. EBNA-5, an Epstein-Barr virus-encoded nuclear antigen, binds to the retinoblastoma and p53 proteins. Proc Natl Acad Sci USA 1993; 90:5455-5459.

68. Henkler F, Waseem N, Golding MH et al. Mutant p53 but not hepatitis B virus X protein is present in hepatitis B virus-related human hepatocellular carcinoma. Cancer Res 1995; 55:6084-6091.

69. Allday MJ, Sinclair A, Parker G et al. Epstein-Barr virus efficiently immortalises human B cells without neutralising the function of p53. EMBO J 1995; 14:1382-1391.

70. Reich NC, Oren M, Levine AJ. Two distinct mechanisms regulate the levels of a cellular tumor antigen, p53. Mol Cell Biol 1983; 3:2143-2150.

71. Hubbert NL, Sedman SA, Schiller JT. Human Papillomavirus type 16 E6 increases the degradation rate of p53 in human keratinocytes. J Virol 1992; 66:6237-6241.

72. Scheffner M, Münger K, Byrne JC et al. The state of the p53 and retinoblastoma genes in human cervical carcinoma cell lines. Proc Natl Acad Sci USA 1991; 88:5523-5527.

73. Scheffner M, Werness BA, Huibregtse JM et al. The E6 oncoprotein encoded by human papillomavirus types 16 and 18 promotes the degradation of p53. Cell 1990; 63:1129-1136.

74. Chowdary DR, Dermody JJ, Jha KK et al. Accumulation of p53 in a mutant cell line defective in the ubiquitin pathway. Mol Cell Biol 1994; 14:1997-2003.

75. Maki C, Huibregtse JM, Howley PM. In vivo ubiquitination and proteasome-mediated degradation of p53. Cancer Res 1996; 56: 2649-2654.

76. Kessis TD, Slebos RJ, Nelson WG et al. Human papillomavirus 16 E6 expression disrupts the p53-mediated cellular response to DNA damage. Proc Natl Acad Sci USA 1993; 90:3988-3992.

77. Foster SA, Demers GW, Etscheid BG et al. The ability of human papillomavirus E6 proteins to target p53 for degradation in vivo correlates with their ability to abrogate actinomycin-D-induced growth arrest. J Virol 1994; 5698-5705.

78. Ciechanover A. The ubiquitin-proteasome proteolytic pathway. Cell 1994; 79:13-21.

79. Hochstrasser M. Ubiquitin, proteasomes, and the regulation of intracellular protein degradation. Curr Opin Cell Biol 1995; 7:215-223.

80. Jentsch S, Schlenker S. Selective protein degradation: a journey's end within the proteasome. Cell 1995; 82:881-884.
81. Scheffner M, Nuber U, Huibregtse JM. Protein ubiquitination involving an E1-E2-E3 enzyme ubiquitin thioester cascade. Nature 1995; 373:81-83.
82. Huibregtse JM, Scheffner M, Beaudenon S et al. A family of proteins structurally and functionally related to the E6-AP ubiquitin-protein ligase. Proc natl Acad Sci USA 1995; 92:2563-2567.
83. Scheffner M., Smith S, Jentsch S. The ubiquitin-conjugation system. In Finley DJ, Peters J-M, Harris R (eds), Plenum Press Inc; Ubiquitin-dependent proteolysis: in press.
84. Scheffner M, Huibregste JM, Vierstra RD et al. The HPV-16 E6 and E6-Ap complex functions as a ubiquitin-protein ligase complex in the ubiquitination of p53. Cell 1993; 75:495-505.
85. Scheffner M., Huibregtse JM, Howley PM. Identification of a human ubiquitin-conjuagting enzyme that mediates the E6-AP-dependent ubiquitination of p53. Proc Natl Acad Sci USA 1994; 91:8797-8801.
86. Nuber U, Schwarz S, Kaiser P et al. Cloning of human conjugating enzymes UbcH6 and UbcH7 (E2-F1) and characterization of their interaction with E6-AP and RSP5. J Biol Chem 1996; 271:2795-2800.
87. Huibregtse JM, Scheffner M, Howley PM. A cellular protein mediates association of p53 with the E6 oncoprotein of human papillomavirus types 16 and 18. EMBO J 1991; 10:4129-4135.
88. Huibregtse JM, Scheffner M, Howley PM. Cloning and expression of the cDNA for E6-AP, a protein that mediates the interaction of the human papillomavirus E6 oncoprotein with p53. Mol Cell Biol 1993; 13:775-784.
89. Sedman SA, Hubbert NA, Vass WC et al. Mutant p53 can substitute for human papillomavirus type 16 E6 in immortalization of human keratinocytes but does not have E6-associated trans-activation or transforming activity. J Virol 1992; 66:4201-4208.
90. Lechner MS, Mack DH, Finicle AB et al. Human papillomavirus E6 proteins bind p53 in vivo and abrogate p53-mediated repression of transcription. EMBO J 1992; 11:3045-3052.
91. Mietz JA, Unger T, Huibregtse JM et al. The transcriptional transactivation function of wild-type p53 is inhibited by SV40 large T-antigen and by HPV-16 oncoprotein. EMBO J 1992; 11:5013-5020.
92. Hoppe-Seyler F, Butz K. Repression of endogenous p53 transactivation function in HeLa cervical carcinoma cells by human papillomavirus type 16 E6, human mdm-2, and mutant p53. J Virol 1993; 67: 3111-3117.
93. Crook T, Fisher C, Masterson PJ et al. Modulation of transcriptional regulatory properties of p53 by HPV E6. Oncogene 1994; 1225-1230.
94. Gu Z, Pim D, Labrecque S et al. DNA damage induced p53 mediated transcription is inhibited by human papillomavirus type 18 E6. Oncogene 1994; 9:629-633.

95. Hoppe-Seyler F, Butz K. Molecular mechanisms of virus-induced carcinogenesis: the interaction of viral factors with cellular tumor suppressor proteins. J Mol Med 1995; 73:529-538.

96. Pim D, Storey A, Thomas M et al. Mutational analysis of HPV-18 E6 identifies domains required for p53 degradation in vitro, abolition of p53 transactivation in vivo and immortalisation of primary BMK cells. Oncogene 1994; 9:1869-1876.

97. Thomas M, Massimi P, Jenkins J et al. HPV-18 E6 mediated inhibition of p53 DNA binding is independent of E6 induced degradation. Oncogene 1995; 10:261-268.

98. White AE, Livanos EM, Tlsty TD. Differential disruption of genomic integrity and cell cycle regulation in normal human fibroblasts by the HPV oncoproteins. Genes Dev 1994; 8:666-677.

99. Reznikoff CA, Belair C, Savelieva E et al. Long-term genome stability and minimal genotypic and phenotypic alterations in human HPV16 E7-, but not E6-, immortalized human uroepithelial cells. Genes Dev 1994; 8:2227-2240.

100. Havre PA, Yuan J, Hedrick L et al. p53 inactivation by HPV16 E6 results in increased mutagenesis in human cells. Cancer Res 1995; 55:4420-4424.

101. Lechner MS, Laimins LA. Inhibition of p53 DNA binding by human papillomavirus E6 proteins. J Virol 1994; 68:4262-4273.

102. Kessis TD, Conolly DC, Hedrick L et al. Expression of HPV16 E6 or E7 increases integration of foreign DNA. Oncogene 1996; 13:427-431.

103. Crook T, Wrede T, Vousden KH. p53 point mutation in HPV negative human cervical carcinoma cell lines. Oncogene 1991; 6:873-875.

104. Fujita M, Inoue M, Tanizawa O et al. Alterations of the p53 gene in primary cervical carcinoma with and without human papillomavirus infection. Cancer Res 1992; 52:5323-5328.

105. Helland A, Holm R, Kristensen G et al. Genetic alterations of the TP53 gene, p53 protein expression and HPV infection in primary cervical carcinomas. J Pathol 1993; 171:105-114.

106. Park DJ, Wilczynski SP, Paquette RL et al. p53 mutations in HPV-negative cervical carcinoma. Oncogene 1994; 9:205-210.

107. Butz K, Shahabeddin L, Geisen C et al. Functional p53 protein in human papillomavirus-positive cancer cells. Oncogene 1995; 10: 927-936.

108. Butz K, Geisen C, Ullmann A et al. Cellular responses of HPV-positive cancer cells to genotoxic anti-cancer agents: Repression of E6/E7-oncogene expression and induction of apoptosis. Int J Cancer 1996; 68:506-513.

109. Dowhanick JJ, McBride AA, Howley PM. Suppression of cellular proliferation by the papillomavirus E2 protein. J Virol 1995; 69:7791-7799.

110. Hwang E-S, Naeger LK, DiMaio D. Activation of the endogenous p53 growth inhibitory pathway in HeLa cervical carcinoma cells by ex-

pression of the bovine papillomavirus E2 gene. Oncogene 1996; 795-803.

111. Haupt Y, Rowan S, Shaulian E et al. Induction of apoptosis in HeLa cells by trans-activation-deficient p53. Genes Dev 1995; 9:2170-2183.

112. Hamada K, Alemany R, Zhang W-W et al. Adenovirus-mediated transfer of a wild-type p53 gene and induction of apoptosis in cervical cancer. Cancer Res 1996; 56:3047-3054.

113. Johnson M, Dimitrov D, Vojta PJ et al. Evidence for a p53-independent pathway for upregulation of SDI1/CIP1/WAF1/p21 RNA in human cells. Mol Carcinogen 1994; 11:59-64.

114. Sheik MS, Li X-S, Chen J-C et al. Mechanisms of regulation of WAF1/CIP1 gene expression in human breast carcinoma: role of p53-dependent and independent signal transduction pathways. Oncogene 1994; 9:3407-3415.

115. Strasser A, Harris AW, Jacks T et al. DNA damage can induce apoptosis in proliferating lymphoid cells via p53-independent mechanisms inhibitable by bcl-2. Cell 1994; 79:329-339.

116. Spitkovsky D, Aengeneyndt F, Braspenning J et al. p53-independent regulation of cervical cancer cells by the papillomavirus E6 oncogene. Oncogene 1996; 13:1027-1036.

117. Storey A, Massimi P, Dawson K et al. Conditional immortalization of primary cells by human papillomavirus type 16 E6 and EJ ras defines an E6 activity in G0/G1 phase which can be substituted for mutations in p53. Oncogene 1995; 11:653-661.

118. Steger G, Pfister H. In vitro expressed HPV 8 E6 protein does not bind p53. Arch Virol 1992; 125:355-360.

119. Kyono T, Hiraiwa A, Ishii S et al. Inhibition of p53-mediated transactivation by E6 of type 1, but not type 5, 8, or 47, human papillomavirus of cutaneous origin. J Virol 1994; 68:4656-4661.

120. Iftner T, Bierfelder S, Csapo Z et al. Involvement of human papillomavirus type 8 E6 and E7 genes in transformation and replication. J Virol 1988; 62:3655-3661.

121. Shay JW, Wright WE, Brasiskyte D et al. E6 of human papillomavirus type 16 can overcome the M1 stage of immortalization in human mammary epithelial cells but not in human fibroblasts. Oncogene 1993; 8:1407-1413.

122. Lamberti C, Morrissey LC, Grossman SR et al. Transcriptional transactivation by the human papillomavirus E6 zinc finger protein. EMBO J 1990; 9:1907-1913.

123. Desaintes C, Hallez S, van Alphen P et al. Transcriptional activation of several heterologous promoters by the E6 protein of human papillomavirus type 16. J Virol 1992; 325:333.

124. Etscheid BG, Foster SA, Galloway DA. The E6 protein of human papillomavirus type 16 functions as a transcriptional repressor in a mechanism independent of the tumor suppressor protein, p53. Virology 1994; 205:583-585.

125. Pan H, Griep AE. Temporally distinct patterns of p53-dependent and p53-independent apoptosis during mouse lens development. Genes Dev 1995; 9:2157-2169.

126. Klingelhutz AJ, Foster SA, McDougall JK. Telomerase activation by the E6 gene product of human papillomavirus type 16. Nature 1996; 380:79-82.

127. Holt SE, Shay JW, Wright WE. Refining the telomere-telomerase hypothesis of aging and cancer. Nature Biotech 1996; 14:836-839.

128. Härle-Bachor C, Boukamp P. Telomerase activity in the regenerative basal layer of the epidermis in human skin and in immortal carcinoma-derived skin keratinocytes. Proc Natl Acad Sci USA 1996; 93:6476-6481.

129. Yasumoto S, Kunimara C, Kikuchi K et al. Telomerase activity in normal human epithelial cells. Oncogene 1996; 13:433-439.

130. Chen JJ, Reid CE, Band V et al. Interaction of papillomavirus E6 oncoproteins with a putative calcium-binding protein. Science 1995; 269:529-531.

131. Keen N, Elston R, Crawford L. Interaction of the E6 protein of human papillomavirus with cellular proteins. Oncogene 1994; 9: 1493-1499.

132. Moran E. Interaction of adenovirus proteins with pRb and p53. FASEB J 1993; 880-885.

133. Debbas M, White E. Wild-type p53 mediates apoptosis by E1A, which is inhibited by E1B. Genes Dev 1993; 7:546-554.

134. Lowe SW, Ruley HE. Stabilization of the p53 tumor suppressor is induced by adenovirus 5 E1A and accompanies apoptosis. Genes Dev 1993; 7:535-545.

135. Putenveettil JA, Frederickson SM, Reznikoff CA. Apoptosis in human papillomavirus16 E7-, but not E6-immortalized human uroepithelial cells. Oncogene 1996; 13:1123-1131.

136. Weinberg RA. The retinoblastoma protein and cell cycle control. Cell 1995; 81:323-330.

137. Demers GW, Halbert CL, Galloway D. Elevated wild-type p53 protein levels in human epithelial cell lines immortalized by the human papillomavirus type 16 E7 gene. Virology 1994; 198:169-174.

138. Androphy EJ. Molecular biology of human papillomavirus infection and oncogenesis. J Invest Dermatol 1994; 103:248-256.

139. von Knebel Doeberitz M, Oltersdorf T, Schwarz E et al. Correlation of modified human papillomavirus early gene expression with altered growth properties in C4-1 cervical carcinoma cells. Cancer Res 1988; 48:3780-3786.

140. von Knebel Doeberitz M, Rittmüller C, zur Hausen H et al. Inhibition of tumorigenicity of C4-1 cervical cancer cells in nude mice by HPV18 E6-E7 antisense RNA. Int J Cancer 1992; 51:831-834.

141. Ham J, Dostatni N, Gauthier J-M et al. The papillomavirus E2 protein: a factor with many talents. Trends Biochem Sci 1991; 16:440-444.
142. McBride AA, Romanczuk H, Howley PM. The papillomavirus E2 regulatory proteins. J Biol Chem 1991; 18411-18414.
143. Pepinsky RB, Androphy EJ, Corina K et al. Specific inhibition of the human papillomavirus E2 trans-activator by intracellular delivery of its repressor. DNA Cell Biol 1994; 13:1011-1019.
144. Storey A, Oates D, Banks L et al. Anti-sense phosphorothiate oligonucleotides have both specific and nonspecific effects on cells containing human papillomavirus type 16. Nucleic Acids Res 1991; 19:4109-4114.
145. Steele C, Sacks PG, Adler-Storthz K et al. Effect on cancer cells of plasmids that express antisense RNA of human papillomavirus type 18. Cancer Res 1992; 52:4706-4711.
146. Steele C, Cowsert LM, Shillitoe EJ. Effects of human papillomavirus type 18-specific antisense oligonucleotides on the transformed phenotype of human carcinoma cell lines. Cancer Res 1993; 53 (Suppl 10):2330-2337.
147. Watanabe S, Kanda T, Yoshiike K. Growth dependence of human papillomavirus 16 DNA-positive cervical cancer cell lines and human papillomavirus 16-transformed human and rat cells on the viral oncoproteins. Jpn J Cancer Res 1993; 84:1043-1049.
148. Hu G, Liu W, Hanania EG et al. Suppression of tumorigenesis by transcription units expressing the antisense E6 and E7 messenger RNA (mRNA) for the transforming proteins of the human papilloma virus and the sense mRNA for the retinoblastoma gene in cervical carcinoma cells. Cancer Gene Ther 1995; 2:263-271.
149. Lu D, Chatterjee S, Brar D et al. Ribozyme-mediated in vitro cleavage of transcripts arising from the major transforming genes of human papillomavirus type 16. Cancer Gene Ther 1994; 1:267-277.
150. Chen Z, Kamath P, Zhang S et al. Effectiveness of three ribozymes for cleavage of an RNA transcript from human papillomavirus type 18. Cancer Gene Ther 1995; 2:263-271.
151. Huibregtse JM, Scheffner M, Howley PM. Localization of the E6-AP regions that direct human papillomavirus E6 binding, association with p53, and ubiquitination of associated proteins. Mol Cell Biol 1993; 13:4918-4927.

CHAPTER 4

E7 Protein

Massimo Tommasino and Pidder Jansen-Dürr

4.1. INTRODUCTION

The first evidence for an involvement of the E7 protein in malignant transformation came from studies on HPV-infected cells. In several cervical carcinoma-derived cell lines, such as SiHa, CaSki and HeLa, the HPV DNA is integrated in the cellular genome as single or multiple copies. This viral DNA integration results in the disruption of several viral genes with consistent preservation of only the early E6 and E7 genes (for more detail see chapter 2). The involvement of the E7 protein in immortalization and transformation of the host cell was confirmed by a number of in vitro assays and by transgenic mouse models. Biochemical studies have provided further evidence that the E7 protein is directly involved in the induction of immortalization and malignant transformation of the host cells. E7 proteins from the "high risk" HPV types have the ability to interact with and alter the function of a number of cellular proteins which play a key role in controlling the proliferative program of the cell.

4.2. DOMAIN STRUCTURE OF THE MOLECULE

The most studied HPV genotypes are those which infect the mucosa of the anogenital tract, and in particular the "high risk"

Papillomaviruses in Human Cancer: The Role of E6 and E7 Oncoproteins,
edited by Massimo Tommasino. © 1997 Landes Bioscience.

HPV type 16. Therefore, the HPV16 E7 will be taken as model and subsequently compared with other E7 proteins from different HPV types.

HPV16 E7 is a small phosphoprotein of 98 amino acids with a predicted isoelectric point of 4.[1] The N-terminal domain (amino acid 1-38) is extremely hydrophilic, while the C-terminal region (amino acid 39-98) is more hydrophobic, despite its predicted molecular weight of about 11 kDa. The HPV16 E7 protein migrates in SDS polyacrylamide electrophoresis as 18-20 kDa molecule. The acidic N-terminal region of E7 appears to be responsible for the aberrant electrophoretic mobility.[2-4]

E7 is structurally and functionally related to other DNA tumor virus proteins, such as the simian virus 40 large tumor antigen (SV40 TAg) and the adenovirus E1A (Ad E1A) protein.[5,6] On the basis of homology with Ad E1A, the HPV16 E7 protein can be divided into three domains: conserved region 1 (CR1), conserved region 2 (CR2) and conserved region 3 (CR3) (Fig. 4.1). Several studies have shown that all three domains are important for the biological activity of the E7 protein. In fact, certain point mutations or small deletions in these regions abrogate its cell transformation activity.

4.2.1. HPV16 E7 CR1 (amino acid 1-20)

CR1 comprises a short amino acid stretch (position 6-15), which is highly conserved in the CR1 of Ad E1A (Fig. 4.1A).[6] Despite the sequence homology between the CR1 of HPV16 E7 and Ad E1A, the two regions appear to have considerable dissimilarities. Ad E1A CR1 has the ability to interact with the cellular protein p300 and with low affinity with the product of the tumor suppressor gene retinoblastoma (pRb), while HPV16 E7 CR1 seems not to have any of these properties.[7-10] Deletions or point mutations in the CR1 of HPV16 E7 lead to a strong reduction in its cellular transformation activity.[11-13] It is not yet clear why an intact CR1 is necessary for E7 transforming activity. Besides, no HPV16 E7 CR1 interacting proteins have been identified so far. The substitution of the histidine in position 2 (H2) with a proline (or aspartic acid) in CR1 of HPV16 E7 abrogates both its transforming activity and its ability to bypass several types of growth arrest.[11-15] A comparison of amino acid sequences of E7 proteins from different HPV genotypes reveals that in position 2 H or R, another positively charged amino acid, are present in about 75% of all E7

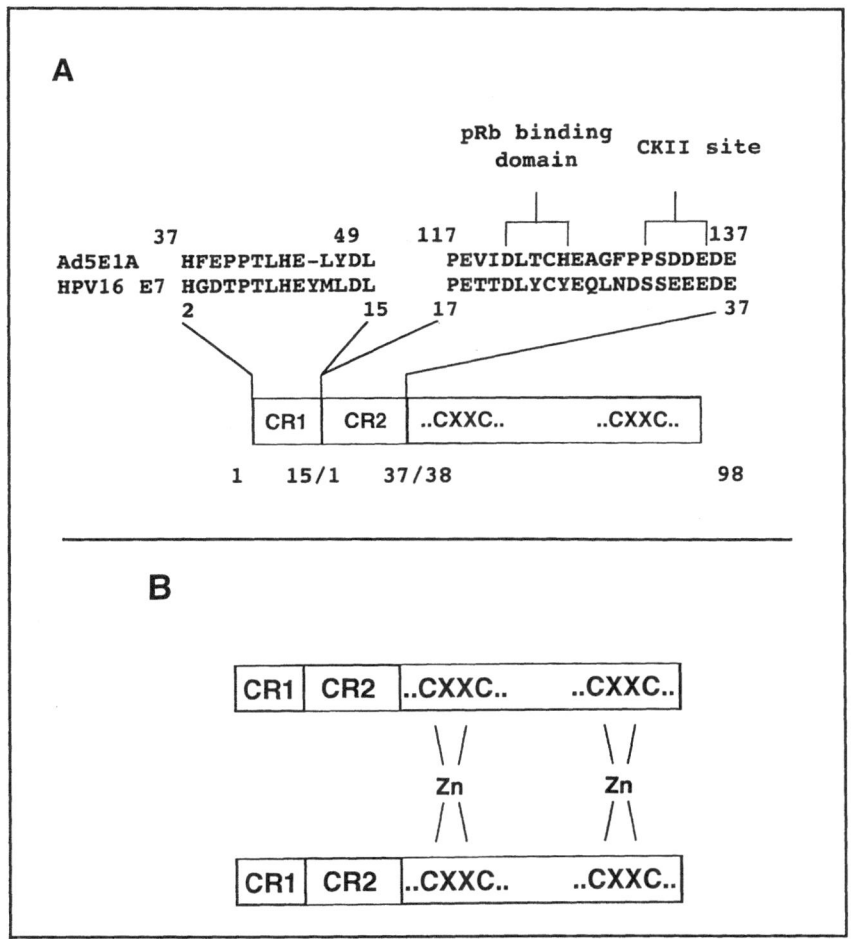

Fig. 4.1. Schematic diagram of HPV 16 E7 protein.
(A) The pRb binding site (LXCXE) is located at position 22-26 and the Casein Kinase II (CKII) phosphorylation site (SS) at position 31-32. The numbers indicate the position of each conserved region (CR 1, 2, and 3). In the upper part of the figure the homology between E7 and E1A in CR1 and CR2 is shown. Bold face characters indicate the conserved amino acids.
(B) Model for dimerization of the E7 protein through CR3.

proteins identified so far, while an aliphatic amino acid (V or I) has been found in the same position in the remaining E7 proteins. However, the presence of an aliphatic or positively charged amino acid in position 2 of E7 protein does not correlate with a particular subgroup of HPV genotypes.

4.2.2. HPV16 E7 CR2 (amino acid 21-38)

CR2 contains the LXCXE domain (amino acid 22-26 in HPV16 E7) which mediates the association with the so-called "pocket proteins", i.e., the tumor suppressor protein pRb and its related proteins, p107 and p130.[16-18] Any mutations in this domain abolish the binding to pRb and abrogate the transformation activity of E7 (see below). The E7 proteins of the "low risk" HPV6 and 11 genotypes have much lower efficiencies in the transformation of primary rodent cells and in the immortalization of human keratinocytes (see section 4.3 for a detailed description of the E7 in vitro transformation assays).[19-20] This correlates with the much lower affinity of HPV6 and 11 E7s for pRb, relative to the "high risk" HPV E7 proteins.[21] Comparison of the pRb binding domains of the "high risk" and "low risk" E7 proteins reveals a characteristic difference of one amino acid, aspartic acid 21 in HPV16 E7 versus glycine 22 in HPV6 and 11 E7 (Fig. 4.2A). Substitution of glycine 22 with an aspartic acid in HPV6 E7 confers greater affinity to bind pRb and the ability to cooperate with activated *ras* oncogene in the transformation of primary rodent cells.[4,22] However, the presence of an aspartic acid in front of the LXCXE domain does not always correlate with the ability to bind pRb with high affinity: the aspartic acid immediately before the LXCXE domain is present in most of all E7 proteins reported, except for HPV6, 7, 11, 13, 29, 40, 44, 55 and 74 (Fig. 4.2A). The E7 proteins from the cutaneous "high risk" HPV types 5 and 8 associate with pRb with much less affinity than HPV16 E7, despite the presence of aspartic acid in the CR2.[23,24] These data indicate that other region(s) of E7, in addition to the LXCXE domain, may mediate the association with pRb. In support of this hypothesis, it was shown that although an HPV16 E7-derived peptide (amino acid 21-29), which

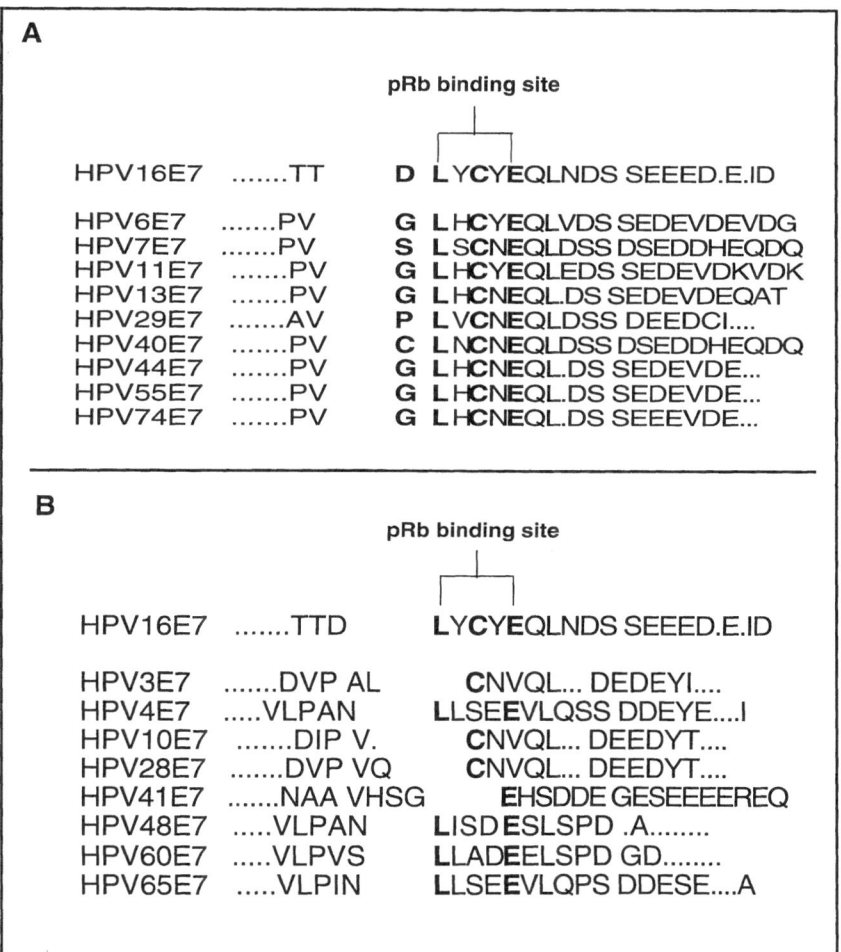

Fig. 4.2. Comparison of "pocket proteins" binding site of E7 proteins from different HPV genotypes.

comprises the "pocket" binding domain, was able to inhibit the E7/pRb interaction, a higher concentration of the peptide was required as compared to the full length E7 protein. This finding indicates that the CR2 is not the only domain involved in the association with Rb1.[25] Furthermore, studies with HPV16 E7 recombinant protein revealed the presence of a low affinity pRb binding site in the C-terminal domain of the molecule.[26] Thus, HPV16 E7 contains two independent pRb binding domains with high and low affinity, which are located in CR2 and CR3, respectively. Although, the LXCXE motif is present in the vast majority of the E7 proteins characterized to date, the proteins encoded by HPV3, 4, 10, 28, 41, 48, 50, 60 and 65 are partially or completely devoid of the "pocket protein" binding consensus motif (Fig. 4.2B). However, it is not yet known if any of these E7 proteins retain the ability to associate with pRb and the related proteins, p107 and p130.

The carboxy-terminal region of CR2 contains a casein kinase II (CKII) site comprising serine or threonine. In the case of HPV16 E7 protein, serines 31 and 32 were shown to be phosphorylated in vitro by CKII.[27] Phosphorylation of the same residues was also observed in vivo.[27] The biological significance of this phosphorylation, however, is not entirely clear, although there is some evidence that this phosphorylation site may be involved in transformation. The "high risk" HPV E7s are better substrates for CKII and changes at this site impair the transforming potential of E7 protein without affecting its association with pRb protein.[12,20,28] Replacement of the two serines by uncharged alanine residues drastically reduced the ability of E7 to cotransform primary cells with *ras* (see section 4.3 for a detailed description of the E7 inv itro transformation assays), whereas replacement by negatively charged aspartic acid at the same positions resulted in a wild-type phenotype.[28] Some E7 proteins lack the T or S residue in the C-terminal part of CR2, although the stretch of negatively charged amino acid is conserved, which could mimic the phosphorylated amino acid residues. Recent data have shown that the HPV16 E7 protein associates with the TATA box binding protein (TBP), and that such interaction is strongly enhanced by phosphorylation of the two serines (see also section 4.5).[29]

4.2.3. HPV16 E7 CR3 (amino acid 39-98)

CR3 of HPV16 E7 has very little homology with Ad E1A CR3, but both domains contain two CXXC motifs involved in zinc binding.[30,31] Using a recombinant E7 protein, it was possible to determine the Zn content of E7 by atomic emission spectroscopy, which revealed a 1:1 molar ratio. X-ray absorption fine structure studies indicated that the Zn is coordinated by four sulphur ligands.[32] Circular dichroism spectroscopy indicated that the E7/Zn interaction results in an increase in the α-helical content and consequent stabilization of a hydrophobic core in the C-terminus of E7.[32] The HPV16 E7 CR3 domain does not resemble any of the known consensus sequences of zinc finger proteins. In particular, its central loop is approximately three times the length of loops contained in most zinc finger structures. Several data suggest that the binding of zinc through the CXXC motif is involved in dimerization of the E7 protein (Fig. 4.1B).[33] Both recombinant bacterially synthesized HPV16 and 18 E7 proteins associate into a higher-order structure, as shown by gel filtration chromatography.[32,33] HPV16 E7 CR3 was able to dimerize in the yeast "two-hybrid system," but only when both of the CXXC motifs were left intact.[34,35] In addition, mutation in one of the two CXXC motifs severely impairs its transformation activity without affecting the ability to bind the pRb protein.[33] Thus, it is possible that dimerization is an important event for the transforming activity of E7. The comparison among the E7 protein sequences reveals that the two CXXC motifs are present in all HPV E7 types, with the exception of the E7 proteins of HPV24, 49 and 54, where one of the CXXC motif is substituted by a CXXXC motif. Interestingly, although the position of the two CXXC motifs can vary in the CR3 E7 sequences, the distance between these motifs is tightly conserved: 29 amino acids in most of the cases and 30 amino acids in HPV4, 41 and 54 E7s. This suggests that the length of the central loop is important for the correct folding of the proteins. In support of this, small deletions in the spacer region between the two CXXC motifs decrease the stability and the transforming activity of the HPV18 E7 protein.[36]

4.3. SUBCELLULAR LOCALIZATION

Early experiments of cellular fractionation of cells derived from the HPV16-positive cervical carcinoma cell line CaSki suggested that E7 is localized exclusively in the cytoplasm.[1] Further studies in monkey COS-1 cells, in which the HPV16 E7 gene was ectopically expressed, confirmed that the majority of the protein upon cellular fractionation was present in the cytoplasm, but it was detectable in both the nuclei and cytoplasm by immunofluorescence staining.[37] The discrepancy between the results obtained by either methods may be due to the release of the E7 protein from nuclei during the disruption of the cellular structure. It is possible that the E7 protein may freely pass through the nuclear pores when the structure of the cell is broken.

Nuclear localization of E7 was also found in BHK-21 cells infected with a recombinant vaccinia virus expressing the HPV16 E7 gene.[38] Kanda et al found that the E7 protein transiently expressed in COS-1 cells could not be detected in the nuclei by indirect immunofluorescence staining with monoclonal antibodies recognizing epitopes in two immunodominant regions of the protein, comprising amino acids 8-22 and 39-54. However, when an anti-E7 polyclonal antibody was used, a clear nuclear staining was observed.[39] These data indicate that two domains of HPV16 E7 protein are masked in the nuclei, probably because they are involved in interactions with other cellular proteins. Apparently the pRb interaction is not responsible for masking of the intracellular E7 protein, since the HPV16 E7 C$\rightarrow$24$\rightarrow$G mutant, which is not able to associate with pRb, was also masked as well as the HPV16 E7 wild type.[39] Similar conclusions were reached when HPV16 E7 protein was fused to the bacterial β-galactosidase (β-gal/E7) and expressed in Rat 3Y1 cells. The chimeric protein appeared to be localized in the nuclei.[40] Deletion analysis of the E7 protein indicated that the domain corresponding to CR2 (amino acid 16-41) was sufficient to target the β-gal/E7 fusion protein into the nucleus. Point mutations of E7 in the LXCXE motif did not alter the nuclear localization of the β-gal/E7 fusion protein, indicating that pRb is not responsible for the translocation of E7 into the nucleus.[40] This is in agreement with data obtained by expressing the HPV16 E7 gene in fission yeast *Schizosaccharomyces pombe*, where none of the "pocket proteins" have been identified and still E7 protein was localized in the nucleus.[41]

Surprisingly, none of the antibodies described above were able to stain E7 in cervical carcinoma cells, such as CaSki or SiHa.[37,39]

Fujikava et al have generated a new series of monoclonal antibodies. One of them was able to specifically stain E7 in CaSki cells, which yielded mainly cytoplasmic staining.[40] Studies by immunofluorescence and electron microscopy immuno-gold staining, using another series of monoclonal antibodies, revealed that a considerable amount of HPV16 E7 protein is localized in the nucleoli of CaSki cells.[42] Greenfield et al showed that E7 is also associated with the nuclear matrix in CaSki and SiHa cells, and that such association is evident only when the cells are subjected to gentle sequential fractionation of the nuclear constituents.[43]

In sum, the data described above clearly show that the E7 protein is localized in several cellular compartments and suggest that the transformation of the host cell induced by E7 requires the alteration of several cellular pathways.

4.4. IMMORTALIZING AND TRANSFORMING ACTIVITIES

4.4.1. Transformation of established rodent cell lines

The E7 proteins of the "high risk" HPV16 and 18 genotypes are able to induce focus formation and growth in soft agar in a variety of established rodent fibroblast lines, e.g., NIH 3T3 and Y31 cells which were tumorigenic in nude mice.[44-46] The E7 proteins of the "low risk" HPV 6 and 11 types can induce similar morphological transformation in immortalized rodent fibroblast, but to a lesser extent than "high risk" HPV E7 proteins.[44,45,47,48] Immortalized rodent fibroblast expressing the "low risk" E7 genes are not able to grow in soft agar, but retain the ability to be tumorigenic in nude mice.[46,49]

The transformation efficiency in rodent fibroblasts of the different E7 proteins does not always correlate with in vivo tumorigenicity of the corresponding virus. The E7 proteins of the "high risk" cutaneous HPV5 and 8 genotypes do not have any transforming activity in established rodent fibroblast, whereas the E7 protein of benign HPV1 genotype can efficiently induce growth in soft agar in NIH 3T3 and C127 cells.[23,24] Since the HPV1 E7 protein binds pRb with greater affinity than HPV5 and 8 E7s, it is conceivable to assume that the transforming activity of E7 proteins in rodent fibroblasts is linked to their efficiency in binding pRb.[23,24,50]

4.4.2. Immortalization and transformation of primary rodent cells

HPV16 and 18 E7 proteins alone are able to immortalize primary rodent cells (rat embryo fibroblast, baby mouse or rat kidney cells),[44,45] but for full malignant transformation the presence of a second oncogene is required. Both E7 proteins cooperate with the activated *ras* gene product, *fos, c-myc* or certain mutant alleles of the cellular p53 gene.[44,51-54] Primary rodent cells transformed by the coexpression of HPV16 E7 and an activated *ras* gene were tumorigenic in syngeneic immunocompetent animals.[51,52] Studies on transformed baby rat kidney cells expressing the HPV16 E7 gene under the control of an inducible promoter, showed that continued expression of the HPV16 E7 gene is necessary for maintenance of the transformed phenotype.[55] The E7 proteins of the "low risk" HPV6 and 11 or skin malignant HPV5 and 8 genotypes can also cooperate with Ha-*ras* to transform primary rodent cells, but at a greatly reduced level compared to HPV16 E7.[23,56,57] Recently, it has also been shown that HPV16 E7 can transform primary rodent cells in cooperation with the product of another viral oncogene, the E5 protein.[58,59] This raised the possibility that the E5 protein my cooperate in vivo with E6 and E7 at the early stage of transformation of the host cell, before the integration of the HPV genome and consequent disruption of the E5 gene, which is frequently observed in human tumors (see also chapter 2).

4.4.3. Immortalization of human primary epithelial cells

The E6 and E7 proteins of HPV16 or 18 cooperate to immortalize primary human keratinocytes, the natural host cell of the virus.[46,60-64] Immortalization can be obtained by expression of the viral oncogenes using either heterologous promoter systems or the native upstream regulatory region (URR).[60-62] The keratinocytes immortalized by E6/E7 are nontumorigenic when injected into nude mice.[63,65] The use of retroviral constructs, expressing either the E6 or E7 gene, demonstrated that the HPV16 E7 protein alone is sufficient to immortalize primary human foreskin epithelial cells, although the immortalization frequency is greatly enhanced by the coexpression of E6.[64] Studies on HPV16 positive cervical carcinoma cell lines, where the expression of E6 and E7 can be modulated by dexamethasone, have clearly shown that the presence of both viral proteins is required to maintain the immortalized and transformed phenotype of the keratinocytes.[66] The in vitro immortalizing activity of the HPV16 E7 protein is not only limited to human keratinocytes. Thus, HPV16 E7 protein has the ability to increase the life span and in some cases to fully immortalize additional human cell types, such as human mammary epithelial cells, primary human aortic and myometrial smooth muscle cells and primary human ovarian surface epithelial cells.[67-71]

The E7 and E6 proteins encoded by the benign HPV types (types 1, 6 and 11) did not show any activity in the immortalization assays with human primary keratinocytes.[19,46] Nevertheless, the HPV6 E7 displayed a weak immortalizing activity, when coexpressed with the "high risk" E6 protein of HPV16 in human keratinocytes.[72]

4.5. CELLULAR PATHWAYS TARGETED BY E7

As a major function of E7 during HPV infection we consider its ability to deregulate controls of cell cycle progression, by favoring the exit of quiescent cells from G_0 and entry into the S phase. This is achieved by E7-driven modulation of cellular gene expression, and by the direct interaction of E7 with cell cycle regulatory proteins (see Tables 4.1 and 4.2).

4.5.1. Cell cycle regulation of mammalian cells

Two fundamental events occur during the cell cycle: the complete and error-free replication of the genome followed by the duplication of the cell. These events are temporally distinct and separated by two phases, G1, before DNA replication and G2, before

Table 4.1. Cellular targets of E7 proteins

	Function	HPV type	Reference
pRb1	regulatory protein of Pol 1, Pol 2 and Pol 3 machinery	6, 11 (low affinity) 1, 16, 18, 33 (high affinity)	16, 21, 50
p107 and p130	regulatory proteins of pol 2 machinery	6, 11 (low affinity) 1, 16, 18, 33 (high affinity)	17, 18, 50
cyclin A/CDK2 complex	cell cycle regulatory kinase complex	1, 16, 18, 33	17, 50, 125
cyclin E/CDK2 complex	cell cycle regulatory kinase complex	16	126
p27	CDK inhibitor	11(low affinity) 16 (high affinity)	95
TBP	pol 2 machinery	11, 16	29
TAF 110	pol 2 machinery	16	117
AP1	transcription factor	16	115

mitosis (Fig. 4.3). Progression through the cell cycle is controlled by several checkpoints, which guarantee the correct succession of cellular events, e.g., mitosis after DNA replication. The cyclin-dependent protein kinases (CDKs) are key enzymes active at various cell cycle checkpoints. CDK activity is regulated by several mechanisms (illustrated in detail in Fig. 4.4) and their activation is an essential requirement for cell cycle progression.

In the absence of viral oncogenes, the exit of resting mammalian cells from the quiescent state (G_0) is triggered by cell type-

Table 4.2. *HPV16 E7 domains involved in interactions with cellular proteins*

HPV16 E7 Domains	Cellular Partner(s)
CR2 (aa 22-26)	pocket proteins (pRb1, p107, p130)
CR2 (aa 31-32)	CKII phosphorylation site, modulation of TBP binding
CR3 (aa 39-98)	p27KIP1
	pRb (low affinity binding site)

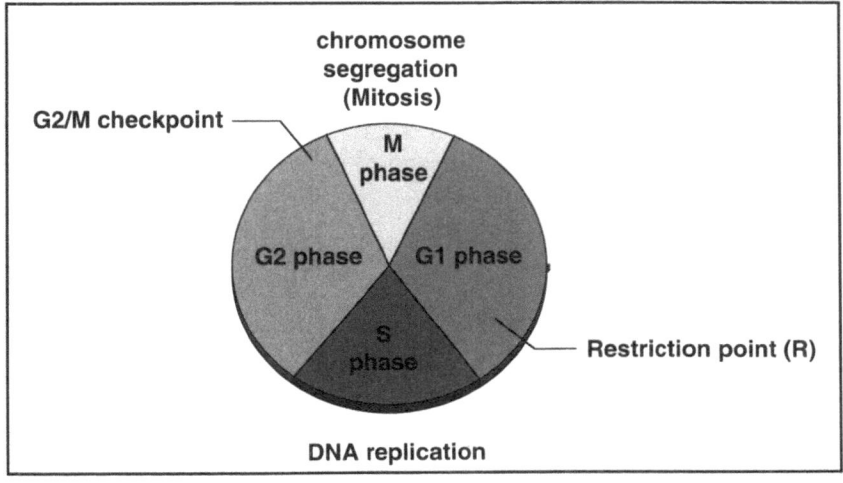

Fig. 4.3. The four phases of the cell cycle (see text for more details).

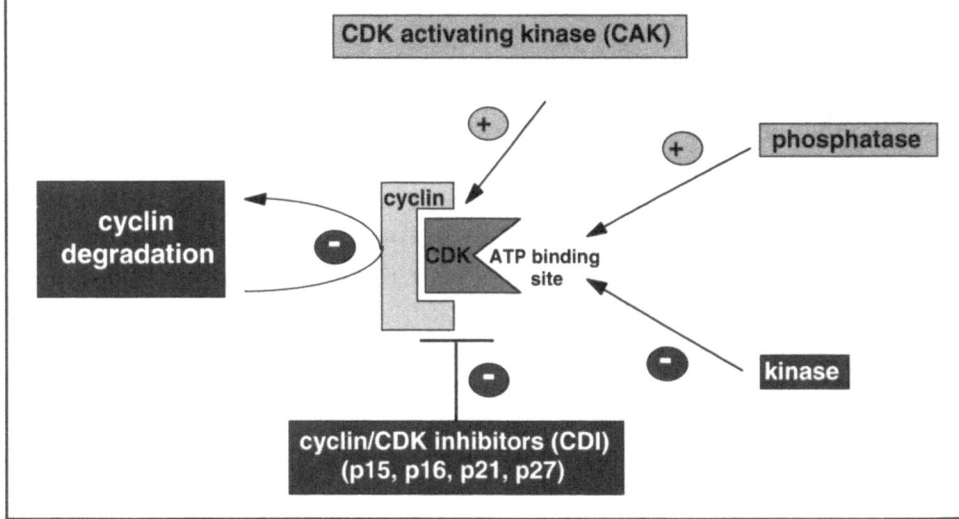

Fig. 4.4. Regulation of cyclin dependent kinase activity. The activity of cyclin/CDK complexes is regulated by several pathways: synthesis or degradation of the regulatory subunit of the kinase complex (cyclin); synthesis and degradation of the CDK inhibitors, p15INK4B, p16INK4a, p21WAF-1 and p27KIP1; phosphorylation by CAK (around amino acid 165) and consequent activation of the CDK subunit; phosphorylation (–) or dephosphorylation (+) of T14 and Y15 at the ATP binding site (for more details see ref. 136).

specific growth factors and their cognate receptors. Growth factor-dependent activation of cyclin D gene expression links receptor-mediated signal transduction to cell cycle control.[73,74] Accumulation of D type cyclins and activation of their associated kinases (CDK4 or CDK6) facilitates progression through G1 by upregulating sthe expression of everal genes, including the cyclin E, E2F-1 and cyclin A genes.[75-80] Proteins of the retinoblastoma gene family pRb, p107 and p130, block cell cycle progression by interacting with and inhibiting cellular transcription factors encoded by the E2F and DP gene families (Fig. 4.5).[81,82] pRb family members are inactivated through multiple phosphorylations by cyclin-dependent kinases, resulting in the release of transcriptionally active E2F/DP heterodimers (Fig. 4.5) (for review see ref. 83). The CDK inhibitors p15INK4B, p16INK4, p21WAF-1 and p27KIP1 arrest cells in G1 by preventing phosphorylation of pRb and related proteins (see Fig. 4.6).[84,85]

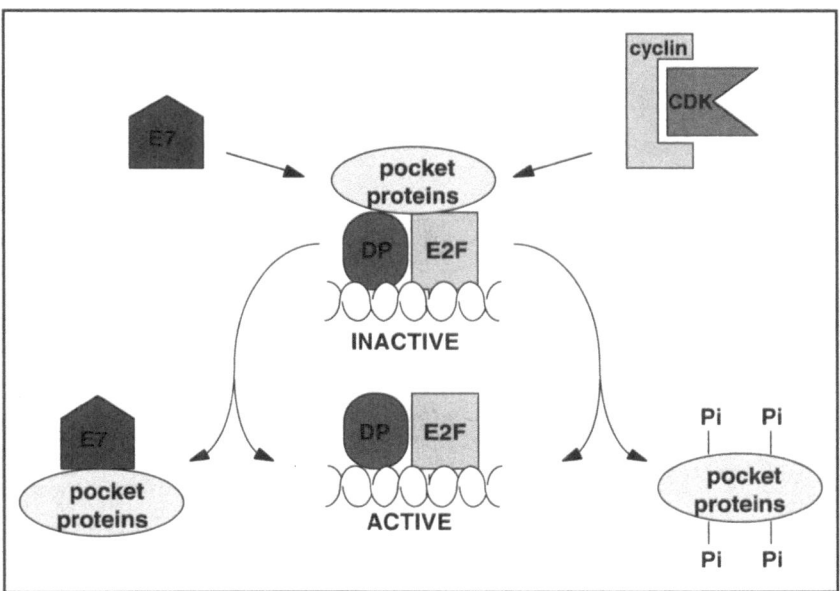

Fig. 4.5. Model for the activation of the E2F/DP transcription factors. In nontrasformed cells the activity of E2F/DP complexes is controlled by phosphorylation and dephosphorylation of the "pocket proteins." Cyclin/CDK complexes are involved in the phosphorylation of the "pocket proteins." E7 binds the hypophosphorylated form of the "pocket proteins" mimicking the effect of phosphorylation by the cyclin/CDK complex.

4.5.2. Cell cycle deregulation by E7

HPV16 E7 allows rodent cells in G1 to enter S phase and increases the proliferative capacity of primary human keratinocytes and of human mammary epithelial cells.[71,86,87] It was shown that E7 can overcome several forms of G1 arrest that are induced by various antiproliferative signals: cells expressing HPV16 E7 continue to enter S phase in the presence of the metabolic inhibitor PALA, when cell adhesion is lost, and when serum growth factors are withdrawn.[11,14,87,88] E7 also keeps cells in S phase when p53 function is induced by either DNA damage or by activation of a temperature-sensitive p53 mutant.[89-92] These results indicate that E7 can override various cell cycle checkpoint controls at the G1/S boundary, and raise the possibility that E7 can neutralize or bypass the inhibitory effects of physiological CDK inhibitors, e.g.,

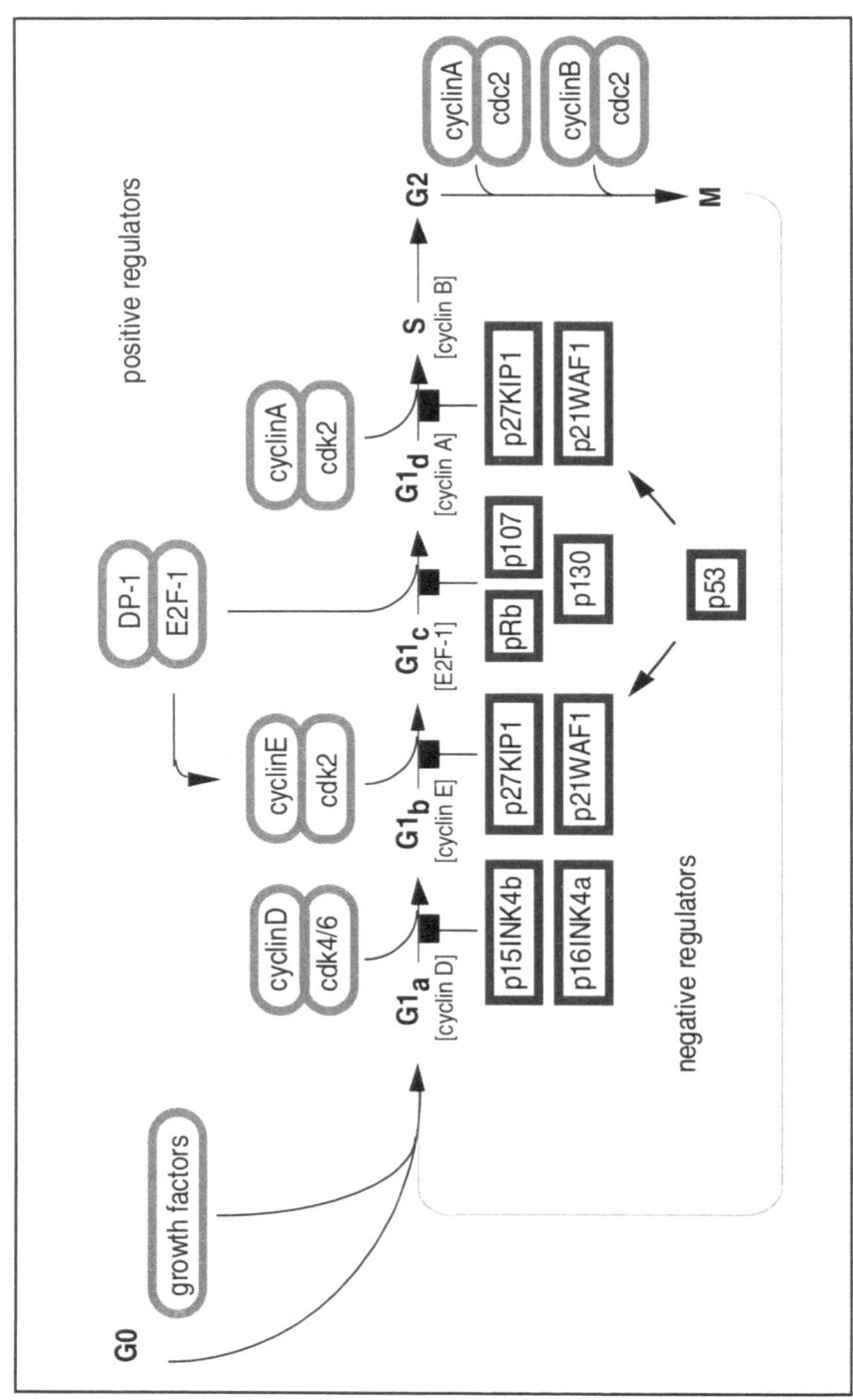

p21WAF1 and P27KIP1, which are induced by p53 and serum with-drawal, respectively (see Fig. 4.6).[85,93,94] In support of this assump-tion, it was shown that HPV16 E7 can inactivate p27KIP1 func-tion in vitro and in transient transfection assays (see ref. 95 and also below).

An increase in the cell proliferation capacity, along with the inhibition of cell differentiation, was also observed when the wild-type HPV16 E7 gene was expressed in the lens of transgenic mice, resulting in microphthalmia and cataracts; in contrast, transgenic mice expressing a CR2 mutant of E7 exhibited normal eyes.[96] These results indicate that E7 promotes S phase entry in vivo, and that the CR2 domain contributes to this phenotype. In agreement with its weak transforming activity in vitro (see section 4.4), the E7 pro-tein of the "low risk" HPV type 6 is unable to override G1 arrest.[87] It was reported that in some cases E7 from high-risk HPV types failed to override G1 arrest induced by p53. Thus, treatment of several HPV-positive cells with genotoxic agents, such as mitomycin C, cisplatin, or UV irradiation, resulted in G1 arrest.[97] Similar re-sults were obtained for rat embryo fibroblasts expressing the HPV16 E7 gene together with activated *ras*.[98] However, the results of the latter experiments are biased by the observation that genotoxic stress abrogates expression of the HPV oncogenes, when expressed from the natural viral transcription unit, at least in one cell line derived from HPV18 positive carcinoma; whether this is a general phenomenon remains to be established.[99] Thus, while it appears possible that the failure of some HPV-positive carcinoma cells to

Fig. 4.6. The role of cyclins, CDKs, CDIs and pRb family members in the control of the mammalian cell cycle (see text for details). The sequential accumulation of cyclin D, cyclin E, E2F-1 and cyclin A is used here to define progression through individual steps of the G1 phase. Positive regulators of cell cycle progression, mainly cyclins and CDKs, are outlined by ovals; negative regulators are boxed. The latter include proteins of the retinoblastoma gene family pRb, p107 and p130, the CDK inhibitors p15INK4B, p16INK4a, p21WAF-1 and p27KIP1 and the p53 tumor suppressor protein. E2F/DP heterodimers are cellular transcription factors that regu-late expression of the cyclin E and A genes; their activity is controlled by pRb family members.

enter S phase upon genotoxic stress is due to insufficient expression of E7 (and potentially also E6), it is clear that the expression of E7 from a strong heterologous promoter is sufficient to override G1 arrest in response to DNA damage.

4.5.3. Interaction of E7 with pRb family proteins; activation of E2F-dependent genes

It was shown that HPV16 E7 associates with the product of the retinoblastoma gene, pRb, and binding results in increased pRb degradation.[16,20,21,24,100,101] By binding to pRb, HPV16 E7 induces the activity of a family of cellular transcription factors, composed of multiple E2F/DP heterodimers (see Fig. 4.5). It was shown that E7 and E2F-1 bind to separate sites on pRb.[102] CR2 is also required for binding of E7 to the pRb-related protein p107 (reviewed in ref. 103). E7 shares with adenovirus E1A the capacity to disrupt the interaction between E2F and pRb,[104,105] an activity that involves both CR2 and CR3 of E7.[26] It was shown that two distinct E2F-p107 complexes exist in different cell cycle phases, and that E7 selectively disrupts the G1-specific complex.[88] Unlike E1A, E7 is unable to disrupt the (S phase-specific) E2F-p107-cyclin A complexes; rather, E7 associates with such complexes.[104,106]

The first E2F-driven gene that was shown to respond to E7 is the E2 gene of adenovirus 5, which contains two tandemly arranged E2F binding sites in the promoter.[107,108] E2F also controls the expression of a series of cellular growth-regulated genes.[81] Of those, the genes encoding B-*myb*, cyclin A and cyclin E were found induced by E7, at least in part via an E2F-dependent mechanism.[14,77,78,109] The importance of E2F-driven genes as targets for the transforming functions of E7 is underscored by the observation that the ability of E7 to activate the cyclin A gene via E2F cosegregates with the transforming potential of E7; furthermore, it was shown that the gene encoding E2F-1 can complement the transforming activity of a CR2 mutant of E7.[14,110] Activation of cyclin E and cyclin A gene expression by E7 in serum-starved cells, i.e., in the absence of cyclin D1, suggests that S phase induction by E7 shortcuts the regulatory cascade of cyclin gene expression, obviating the need for growth factor-induced accumulation of cyclin D1 and its associated kinase.[14] Indeed, cyclin D1-associated kinase was shown to be dispensable in various cell types in which pRb function is compromised, either by mutation of the Rb gene or by the expression of a pRb-inactivating viral gene.[111] In support of this interpretation, it was shown that cyclin D1 overexpression, a characteristic feature of many human tumors, is not observed in HPV-positive cervical cancer.[112]

Accumulating evidence suggests that pRb is a major regulator of transcription by all three eukaryotic RNA polymerases. pRb binds to the pol I activator upstream binding factor (UBF) and represses pol I transcription.[113] Recently it was also shown that the activity of RNA polymerase III is negatively regulated by pRb.[114] At present it is unknown if E7 can affect the ability of pRb to repress expression of genes transcribed by either pol I or pol III, although it was shown that an E7 CR2 peptide, containing the LXCXE motif, was able to counteract the inhibitory effect of pRb on pol I in vitro.[113]

4.5.4. Modulation of Rb-independent transcriptional pathways by E7

Besides pRb and its relatives, other transcriptional regulatory proteins were described as targets for E7. Thus, HPV16 E7 interacts with the AP1 family of transcription factors and this results in trans-activation of AP-1 driven reporter genes.[115] Similarly, it was shown that the HPV16 E7 protein can complement certain mutants of Ad 5 E1A in the trans-activation of adenovirus early genes, an activity which may involve the transcription factors ATF and Oct-1.[116] These findings suggest that E7 may target cellular (and potentially viral) genes regulated by these transcription factors. Like Ad5 E1A, HPV16 E7 can bind to the TATA box binding protein (TBP) and the TBP-associated factor TAF110, indicating that E7 may influence the activity of these general transcription factors.[29,117] However, the precise role of these interactions remains to be determined.

The ability of E7 to bind pRb is not necessarily required for the activation of transcription, even in the case of the adenovirus E2 gene promoter. Thus, it was shown that the E7 proteins encoded by the low-risk viruses HPV6 and HPV11 can also trans-activate the Ad5 E2 promoter.[118] In the case of the low-risk virus HPV1, the evidence is conflicting: while it was reported that HPV1 E7 cannot trans-activate the E2 promoter and is unable to relieve the E2F-1 transcription factor from repression by pRb, Ibaraki et al reported that the Ad5 E2 promoter is stimulated to a similar extent by the E7 genes of HPV1 and HPV16.[24,50,119] While the reason for this apparent contradiction is unclear, it appears possible that, as was found for trans-activation of E2 by adenovirus E1A, trans-activation of the E2 promoter by E7 may involve several promoter elements, and part of the trans-activation potential of E7 may be independent of the CR2 domain.[120]

It was shown that a mutation in CR1 decreases trans-activation of the adenovirus E2 gene by HPV16 E7; furthermore, the carboxy-terminus of HPV16 E7 is sufficient for trans-activation of the adenovirus E2 promoter in peptide microinjection studies.[13,121] These results suggest that distinct domains of E7 can independently activate the E2 promoter, and that additional promoter elements besides the E2F site may be targets for E7. The recent demonstration of transcriptional activation domains in CR1 and CR3 of HPV16 E7, which are both independent of E2F/pRb, supports this concept.[35] Interestingly, an equivalent activity was not found in the E7 protein encoded by the "low risk" HPV11.[35] Additional evidence suggests that E7 has the potential to stimulate transcription independent of E2F. Thus, it was shown that HPV16 E7 stimulates expression of the *c-fos* gene via a cyclic AMP response element in the *c-fos* promoter. Mutations in either CR1 or CR2 reduce trans-activation of the *c-fos* gene by E7, and E7 of the "low risk" HPV6 is much less active in activating the *c-fos* promoter than HPV16 E7.[12,21,23] These data indicate that E2F-independent transcriptional activation may represent a specific function of HPV16 E7, which is not conserved in several low-risk HPV types. HPV16 E7 was also shown to activate, in a CR2-dependent manner, transcription of the gene encoding a 100 kDa protein related to heat shock proteins, and it was reported that the smooth-muscle α-actin gene is repressed by HPV16 E7; however, the significance of these findings for cell transformation remains to be demonstrated.[12,24,31]

4.5.5. Physical interaction of E7 with cell cycle regulatory proteins

HPV16 E7 associates with cyclin A and CDK2 through its CR2 domain.[17,125] More recently, it was found that HPV16 E7also binds to cyclin E in a complex with CDK2 and p107.[126] It is possible that the interaction of E7 with cyclins is mediated by p107, which is known to associate with both cyclins A and E.[127] The efficiency of binding to cyclin A appears to correlate with the transforming capacity of the E7 oncoproteins of several human papillomaviruses.[50] However, E7 of the benign type HPV1 binds cyclin A with similar affinity as the "high-risk" E7 proteins.[50] It was reported that expression of HPV16 E7 in normal human fibroblasts dissociates p21WAF1 and proliferating cell nuclear antigen (PCNA) from the quaternary cyclin-CDK complexes and causes selective disruption of the cyclin D-CDK4 complex, which is replaced by a CDK4-p16INK4 complex; in contrast, E7 from a low-risk HPV did not have any effect on these complexes.[128] Disruption of cyclin D1/CDK4 complexes may be caused by E7 only indirectly, i.e., through induction of p16INK4 gene expression.[129] Since the CXXC motifs in CR3 are essential for the immortalization of human keratinocytes, the zinc finger may mediate the interaction of E7 with specific cellular proteins involved in immortalization.[33,130] In line with this assumption, it was shown that the CDK inhibitor p27KIP1 binds to E7 via the C-terminal domain, and E7 binding results in functional inactivation of p27KIP1; this observation may provide a molecular basis for the reported increase of cyclin E/CDK2 kinase activity in E7-expressing cells.[14,95] These findings suggest that cyclin D1-associated kinase is repressed upon expression of E7 while cyclin E-associated kinase becomes activated, in keeping with the observation that while cyclin D1 is dispensable for S-phase entry of HPV-transformed cells, cyclin E is still required.[111,131]

4.6. E7 PROTEIN AND APOPTOSIS

HPV16 E7 has a dual activity being able to induce proliferation and apoptosis. Studies in normal human fibroblasts (NHF) have shown that the expression of HPV16 E7, without the E6 gene, resulted in a cytocidal response.[87] This observation was confirmed by several studies on transgenic mice specifically expressing E7 in the eyes. Transgenic mice expressing HPV16 E7 gene under the control of the interstitial retinol-binding protein promoter display several abnormalities in the eyes.[132] Thus, the retinas of these mice degenerate due to photoreceptor cell death, and this occurs at the same time of development in which photoreceptors would undergo terminal differentiation in wild-type mice. Further analysis clearly demonstrated that the dying cells exhibit the typical molecular and ultrastructural features of apoptosis.[132] Analogous data were obtained when the E7 gene was expressed under the control of the murine αA crystallin promoter in the ocular lens. Here, E7 inhibited lens fiber cell differentiation, inducing proliferation and apoptosis.[96] The integrity of the pRb binding domain of E7 was essential for the induction of proliferation and apoptosis. Transgenic mice expressing an E7 mutant, in which four of the six amino acids involved in pRb binding had been deleted, did not show any abnormality.[96] Studies in p53 null mice indicated that E7 can induce apoptosis also in the absence of p53, although to a lower extent. Thus, E7-induced apoptosis is mediated by p53-dependent and independent pathways.[133]

Recent studies have shown that human uroepithelial cells immortalized by HPV16 E7 have an enhanced apoptotic response upon exposure to γ radiation compared to normal human uroepithelial cells.[134] In these cells, the HPV16 E7-induced apoptosis appears to be mediated entirely by wt p53. In fact, HPV16 E7 immortalized human uroepithelial cells which contain a mutated form of p53, are incapable to enter apoptosis upon exposure to γ radiation.[134]

The physiological meaning of E7 induced apoptosis is not clear. The decision of the E7 expressing cells to undergo apoptosis may represent a cellular defense reaction elicited by the loss of cell cycle control. HPV16 E6 protein, by inducing the degradation of p53 (see chapter 3), counteracts the activity of E7 in inducing apotosis.[96] Therefore, in this scenario, both viral proteins are required to promote immortalization of the host cell.

Despite the fact that both the E6 and E7 genes are transcribed in most if not all HPV-positive lesions, apoptosis has been observed in benign, premalignant and malignant HPV-positive lesions, suggesting that the balance between proliferation and apoptosis is not exclusively regulated by the viral genes.[135] Only 10% of the "high risk" HPV infections progress to the development of malignant lesion. Thus, it is conceivable that such progression is determined by the balance between proliferation and apoptosis. Cells which can cycle many times without entering programmed cell death have a much higher probability of accumulating chromosomal alterations, and becoming a malignant lesion.

4.7. CONCLUSION

HPVs infect epithelial cells and, as these are largely quiescent, have developed methods of moving the cells from G0 into the cell cycle in the absence of external growth stimulatory signals. This is achieved mainly by the E7 activities in cooperation with the E6 protein, and probably to a lesser extent by the E5 protein. The comparison of the in vitro functional activities of the E7 protein with proteins encoded by other DNA tumor viruses, such as SV40 large TAg and adenovirus E1A, was extremely helpful for the understanding of its biological activities in interfering with cellular growth control. Although the adenovirus E1A protein shares most of the known properties of the HPV E7 protein and is even more efficient than HPV E7 in a number of in vitro activities, only HPVs are associated with malignant human tumors. The ability of these DNA tumor viruses to induce malignant lesions in vivo could be due to differences in the capacity of neutralizing the immune host defenses, or alternatively, to subtle differences in the type of transformation produced.

Another discrepancy is raised when the E7 proteins of the different HPV genotypes are compared. Several studies have tried to correlate particular in vitro properties of HPV E7 proteins with the in vivo transforming efficiencies of the corresponding viruses.

The affinity of the different E7 proteins in binding pRb correlates well with the transforming activity of the HPV genotype as long as we restrict our analysis to the mucosal HPV group. However, if cutaneous HPV1 is included in the comparison the in vitro evidence would suggest that HPV 1 should be a "high risk" type, rather than a very low or no risk type as it is known to be. The issue becomes even more complex if we extend the analysis to other E7 proteins. For instance, HPV41E7 lacks the pRb high affinity binding site in CR2, nevertheless this HPV type is clearly associated in vivo with cutaneous malignant lesions.

It is conceivable that the E7 protein interacts with additional cellular proteins and that such interactions most likely also involve the transforming domains CR1 and CR3, for which the biochemical action remains largely unclear. Studies on the E7 protein contributed to our understanding of the role of several host proteins in controlling fundamental cellular events. Further studies on HPV and in particular on E6 and E7 will allow us to identify and better understand additional cellular pathways involved in the regulation and safeguard of the proliferation and differentiation of a normal cell.

REFERENCES

1. Smotkin D, Wettstein FO. The major human papillomavirus protein in cervical cancers is a cytoplasmic phosphoprotein. J Virol 1987; 61:1686-9.
2. Armstrong DJ, Roman A. Mutagenesis of human papillomavirus types 6 and 16 E7 open reading frames alters the electrophoretic mobility of the expressed proteins. J Gen Virol 1992; 73:1275-9.
3. Armstrong DJ, Roman A. The anomalous electrophoretic behavior of the human papillomavirus type 16 E7 protein is due to the high content of acidic amino acid residues. Biochem Biophys Res Commun 1993; 192:1380-7.
4. Sang BC, Barbosa MS. Single amino acid substitutions in "low-risk" human papillomavirus (HPV) type 6 E7 protein enhance features characteristic of the "high-risk" HPV E7 oncoproteins. Proc Natl Acad Sci USA 1992; 89:8063-7.
5. Vousden KH, Jat PS. Functional similarity between HPV16E7, SV40 large T and adenovirus E1a proteins. Oncogene 1989; 4:153-8.

6. Phelps WC, Yee CL, Münger K et al. The human papillomavirus type 16 E7 gene encodes transactivation and transformation functions similar to those of adenovirus E1A. Cell 1988; 53:539-547.

7. Davies RC, Vousden KH. Functional analysis of human papillomavirus type 16 E7 by complementation with adenovirus E1A mutants. J Gen Virol 1992; 73:2135-9.

8. Ikeda MA, Nevins JR. Identification of distinct roles for separate E1A domains in disruption of E2F complexes. Mol and Cell Biol 1993; 13:7029-7035.

9. Fattaey AR, Harlow E, Helin K. Independent regions of adenovirus E1A are required for binding to and dissociation of E2F-protein complexes. Mol and Cell Biol 1993; 13:7267-7277.

10. Whyte P, Williamson NW, Harlow E. Cellular targets for transformation by the adenovirus E1A proteins. Cell 1989; 56:67-75.

11. Banks L, Edmonds C, Vousden K. Ability of the HPV16 E7 protein to bind RB and induce DNA synthesis is not sufficient for efficient transforming activity in NIH3T3 cells. Oncogene 1990; 5:1383-1389.

12. Phelps WC, Münger K, Yee CL et al. Structure-function analysis of the human papillomavirus type 16 E7 oncoprotein. J Virol 1992; 66:2418-2427.

13. Watanabe S, Kanda T, Sato H et al. Mutational analysis of human papillomavirus type 16 E7 functions. J Virol 1990; 64:207-214.

14. Zerfass K, Schulze A, Spitkovsky D et al. Sequential activation of cyclin E and cyclin A gene expression by HPV-16 E7 through sequences necessary for transformation. J Virol 1995; 69:6389-6399.

15. Demers WG, Espling E, Beth Harry J et al. Abrogation of growth arrest signals by human papillomavirus type 16 E7 is mediated by sequences required for transformation. J Virology 1996; 70:6862-6869.

16. Dyson N, Howley PM, Munger K et al. The human papilloma virus-16 E7 oncoprotein is able to bind to the retinoblastoma gene product. Science 1989; 243:934-937.

17. Davies R, Hicks R, Crook T et al. Human papillomavirus type-16 E7 associates with a histone H1 kinase and with p107 through sequences necessary for transformation. J Virol 1993; 67:2521-2528.

18. Hu TH, Ferril SC, Snider AM et al. In vivo analysis of HPV E7 protein association with pRb, p107 and p130. Int J Oncol 1995; 6:167-174.

19. Woodworth CD, Doniger J, DiPaolo JA. Immortalisation of human foreskin keratinocytes by various human papillomavirus DNAs corresponds to their association with cervical carcinoma. J Virol 1989; 63:159-64.

20. Barbosa MS, Edmonds C, Fisher C et al. The region of the HPV E7 oncoprotein homologous to adenovirus E1a and Sv40 large T antigen contains separate domains for Rb binding and casein kinase II phosphorylation. Embo J 1990; 9:153-60.

21. Munger K, Werness BA, Dyson N et al. Complex formation of human papillomavirus E7 proteins with the retinoblastoma tumor suppressor gene product. Embo J 1989; 8:4099-105.
22. Heck DV, Yee CL, Howley PM et al. Efficiency of binding the retinoblastoma protein correlates with the transforming capacity of the E7 oncoproteins of the human papillomaviruses. Proc Natl Acad Sci USA 1992; 89:4442-6.
23. Yamashita T, Segawa K, Fujinaga Y et al. Biological and biochemical activity of E7 genes of the cutaneous human papillomavirus type 5 and 8. Oncogene 1993; 8:2433-41.
24. Schmitt A, Harry JB, Rapp B et al. Comparison of the properties of the E6 and E7 genes of low- and high-risk cutaneous papillomaviruses reveals strongly transforming and high Rb-binding activity for the E7 protein of the low-risk human papillomavirus type 1. J Virol 1994; 68:7051-7059.
25. Jones RE, Wegrzyn RJ, Patrick DR et al. Identification of HPV-16 E7 peptides that are potent antagonists of E7 binding to the retinoblastoma suppressor protein. J Biol Chem 1990; 265:12782-5.
26. Patrick DR, Oliff A, Heimbrook DC. Identification of a novel retinoblastoma gene product binding site on human papillomavirus type 16 E7 protein. J Biol Chem 1994; 269:6842-6850.
27. Firzlaff JM, Galloway DA, Eisenman RN et al. The E7 protein of human papillomavirus type 16 is phosphorylated by casein kinase II. New Biol 1989; 1:44-53.
28. Firzlaff JM, Luscher B, Eisenman RN. Negative charge at the casein kinase II phosphorylation site is important for transformation but not for Rb protein binding by the E7 protein of human papillomavirus type 16. Proc Natl Acad Sci USA 1991; 88:5187-91.
29. Massimi P, Pim D, Storey A et al. HPV-16 E7 and adenovirus E1a complex formation with TATA box binding protein is enhanced by casein kinase II phosphorylation. Oncogene 1996; 12:2325-2330.
30. Culp JS, Webster LC, Friedman DJ et al. The 289-amino acid E1A protein of adenovirus binds zinc in a region that is important for trans-activation. Proc Natl Acad Sci USA 1988; 85:6450-4.
31. Barbosa MS, Lowy DR, Schiller JT. Papillomavirus polypeptides E6 and E7 are zinc-binding proteins. J Virol 1989; 63:1404-7.
32. Patrick DR, Zhang K, Defeo JD et al. Characterization of functional HPV-16 E7 protein produced in *Escherichia coli.* J Biol Chem 1992; 267:6910-5.
33. McIntyre MC, Frattini MG, Grossman SR et al. Human papillomavirus type-18 E7 protein requires intact Cys-X-X-Cys motifs for zinc binding, dimerization, and transformation but not for Rb binding. J Virol 1993; 67:3142-3150.
34. Clemens KE, Brent R, Gyuris J et al. Dimerization of the human papillomavirus E7 oncoprotein in vivo. Virology 1995; 214:289-293.

35. Zwerschke W, Joswig S, Jansen-Dürr P. Identification of domains required for transcriptional activation and protein dimerization in the human papillomavirus type-16 E7 protein. Oncogene 1996; 12:213-220.

36. Watanabe S, Sato H, Furuno A et al. Changing the spacing between metal-binding motifs decreases stability and transforming activity of the human papillomavirus type 18 E7 oncoprotein. Virology 1992; 190:872-5.

37. Sato H, Watanabe S, Furuno A et al. Human papillomavirus type 16 E7 protein expressed in *Escherichia coli* and monkey COS-1 cells: immunofluorescence detection of the nuclear E7 protein. Virology 1989; 170:311-5.

38. Meneguzzi G, Cerni C, Kieny MP et al. Immunization against human papillomavirus type 16 tumor cells with recombinant vaccinia viruses expressing E6 and E7. Virology 1991; 181:62-9.

39. Kanda T, Zanma S, Watanabe S et al. Two immunodominant regions of the human papillomavirus type 16 E7 protein are masked in the nuclei of monkey COS-1 cells. Virology 1991; 182:723-31.

40. Fujikawa K, Furuse M, Uwabe KI et al. Nuclear localization and transforming activity of human papillomavirus type 16 E7-β-galactosidase fusion protein: Characterization of the nuclear localization sequence. Virology 1994; 204:789-793.

41. Tommasino M, Contorni M, Scarlato V et al. Synthesis, phosphorylation, and nuclear localization of human papillomavirus E7 protein in Schizosaccharomyces pombe. Gene 1990; 93:265-70.

42. Zatsepina O, Braspenning J, Robberson D et al. The human papillomavirus type E7 protein is associated with the nucleolus in mammalian and yeast cells. Oncogene 1997;14:1137-1145.

43. Greenfield I, Nickerson J, Penman S et al. Human papillomavirus 16 E7 protein is associated with the nuclear matrix. Proc Natl Acad Sci USA 1991; 88:11217-21.

44. Bedell MA, Jones KH, Grossman SR et al. Identification of human papillomavirus type 18 transforming genes in immortalized and primary cells. J Virol 1989; 63:1247-55.

45. Kanda T, Watanabe S, Yoshiike K. Immortalization of primary rat cells by human papillomavirus type 16 subgenomic DNA fragments controlled by the SV40 promoter. Virology 1988; 165:321-5.

46. Barbosa MS, Vass WC, Lowy DR et al. In vitro biological activities of the E6 and E7 genes vary among human papillomaviruses of different oncogenic potential. J Virol 1991; 65:292-8.

47. Le JY, Defendi V. A viral-cellular junction fragment from a human papillomavirus type 16-positive tumor is competent in transformation of NIH 3T3 cells. J Virol 1988; 62:4420-6.

48. Yutsudo M, Okamoto Y, Hakura A. Functional dissociation of transforming genes of human papillomavirus type 16. Virology 1988; 166:594-7.

49. Tsao YP, Chu TY, Chen TM et al. Effects of E5a and E7 genes of human papillomavirus type 11 on immortalized human epidermal keratinocytes and NIH 3T3 cells. Arch Virol 1994; 138:177-85.
50. Ciccolini F, Di Pasquale G, Carlotti F et al. Functional studies of E7 proteins from different HPV types. Oncogene 1994; 9:2342-2348.
51. Matlashewski G, Schneider J, Banks L et al. Human papillomavirus type 16 DNA cooperates with activated ras in transforming primary cells. Embo J 1987; 6:1741-6.
52. Storey A, Pim D, Murray A et al. Comparison of the in vitro transforming activities of human papillomavirus types. Embo J 1988; 7:1815-20.
53. Peacock JW, Matlashewski GJ, Benchimol S. Synergism between pairs of immortalizing genes in transformation assays of rat embryo fibroblasts. Oncogene 1990; 5:1769-74.
54. Crook T, Fisher C, Vousden KH. Modulation of immortalizing properties of human papillomavirus type 16 E7 by p53 expression. J Virol 1991; 65:505-510.
55. Crook T, Storey A, Almond N et al. Human papillomavirus type 16 cooperates with activated ras ans fos oncogenes in the hormone dependent transformation of primary mouse cells. Proc Natl Acad Sci USA 1988; 85:8820-8824.
56. Storey A, Osborn K, Crawford L. Co-transformation by human papillomavirus types 6 and 11. J Gen Virol 1990; 71:165-71.
57. Nishikawa T, Yamashita T, Yamada T et al. Tumorigenic transformation of primary rat embryonal fibroblasts by human papillomavirus type 8 E7 gene in collaboration with the activated H-ras gene. Jpn J Cancer Res 1991; 82:1340-3.
58. Bouvard V, Matlashewski G, Gu ZM et al. The human papillomavirus type 16 E5 gene cooperates with the E7 gene to stimulate proliferation of primary cells and increases viral gene expression. Virology 1994; 203:73-80.
59. Valle GF, Banks L. The human papillomavirus (HPV)-6 and HPV-16 E5 proteins cooperate with HPV-16 E7 in the transformation of primary rodent cells. J Gen Virol 1995; 76:1239-45.
60. Kaur P, McDougall JK. Characterization of primary human keratinocytes transformed by human papillomavirus type 18. J Virol 1988; 62:1917-24.
61. Hawley NP, Vousden KH, Hubbert NL et al. HPV16 E6 and E7 proteins cooperate to immortalize human foreskin keratinocytes. Embo J 1989; 8:3905-10.
62. Munger K, Phelps WC, Bubb V et al. The E6 and E7 genes of the human papillomavirus type 16 together are necessary and sufficient for transformation of primary human keratinocytes. J Virol 1989; 63:4417-21.

63. Hudson JB, Bedell M A, McCance D J et al. Immortalization and altered differentiation of human keratinocytes in vitro by the E6 and E7 open reading frames of human papillomavirus type 18. J Virol 1990; 64:519-526.

64. Halbert CL, Demers GW, Galloway DA. The E7 gene of human papillomavirus type 16 is sufficient for immortalization of human epithelial cells. J Virol 1991; 65:473-8.

65. Barbosa MS, Schlegel R. The E6 and E7 genes of HPV-18 are sufficient for inducing two-stage in vitro transformation of human keratinocytes. Oncogene 1989; 4:1529-32.

66. von Knebel Doeberitz M, Rittmüller C, Aengenyndt F et al. Reversible repression of papillomavirus oncogene expression in cervical carcinoma cells: consequences for the phenotype and E6-p53 and E7-pRB interactions. J Virol 1994; 68:2811-2821.

67. Perez RN, Halbert CL, Smith PP et al. Immortalization of primary human smooth muscle cells. Proc Natl Acad Sci USA 1992; 89:1224-8.

68. Conroy SC, Hart CE, Perez RN et al. Characterization of human aortic smooth muscle cells expressing HPV16 E6 and E7 open reading frames. Am J Pathol 1995; 147:753-62.

69. Wazer DE, Liu XL, Chu Q et al. Immortalization of distinct human mammary epithelial cell types by human papillomavirus 16 E6 or E7. Proc Natl Acad Sci USA 1995; 92:3687-91.

70. Tsao SW, Mok SC, Fey EG et al. Characterization of human ovarian surface epithelial cells immortalized by human papilloma viral oncogenes (HPV-E6E7 ORFs). Exp Cell Res 1995; 218:499-507.

71. Foster SA, Galloway DA. Human papillomavirus type 16 E7 alleviates a proliferation block in early passage human mammary epithelial cells. Oncogene 1996; 12:1773-9.

72. Halbert CL, Demers GW, Galloway DA. The E6 and E7 genes of human papillomavirus type 6 have weak immortalizing activity in human epithelial cells. J Virol 1992; 66:2125-34.

73. Matsushime H, Roussel MF, Ashmun RA et al. Colony-stimulating factor 1 regulates novel cyclins during the G1 phase of the cell cycle. Cell 1991; 65:701-713.

74. Won K-A, Xiong Y, Beach D et al. Growth-regulated expression of D-type cyclin genes in human diploid fibroblasts. Proc Natl Acad Sci USA 1992; 89:9910-9914.

75. Johnson DG, Ohtani K, Nevins JR. Autoregulatory control of E2F1 expression in response to positive and negative regulators of cell cycle progression. Genes & Dev 1994; 8:1514-1525.

76. Schulze A, Zerfaß K, Spitkovsky D et al. Activation of the E2F transcription factor by cyclin D1 is blocked by p16INK4, the product of the putative tumor suppressor gene MTS1. Oncogene 1994; 9: 3475-3482.

77. Botz J, Zerfass-Thome K, Spitkovsky D et al. Cell cycle regulation of the murine cyclin E gene depends on an E2F binding site in the promoter. Mol Cell Biol 1996; 16:3401-3409.
78. Ohtani K, Degregori J, Nevins JR. Regulation of the cyclin E gene by transcription factor E2F1. Proc Natl Acad Sci USA 1995; 92: 12146-12150.
79. Schulze A, Zerfaß K, Spitkovsky D et al. Cell cycle regulation of cyclin A gene transcription is mediated by a variant E2F binding site. Proc Natl Acad Sci USA 1995; 92:11264-11268.
80. Sherr CJ. Cancer cell cycles. Science 1996; 274:1672-1677.
81. La Thangue NB. Drtf1/E2F—An expanding family of heterodimeric transcription factors implicated in cell-cycle control. Trends Biochem Sci 1994; 19:108-114.
82. Lathangue NB. E2F and the molecular mechanisms of early cell-cycle control. Biochemical Society Transactions 1996; 24:54-59.
83. Beijersbergen RL, Bernards R. Cell cycle regulation by the retinoblastoma family of growth inhibitory proteins. Biochimica et Biophysica Acta—Reviews on Cancer 1996; 1287:103-120.
84. Hunter T, Pines J. Cyclins and cancer 2. Cyclin D and CDK inhibitors come of age. Cell 1994; 79:573-582.
85. El-Deiry WS, Tokino T, Velculescu VE et al. WAF1, a potential mediator of p53 tumor suppression. Cell 1993; 75:817-825.
86. Banks L, Barnett SC, Crook T. HPV-16 E7 functions at the G1 to S phase transition in the cell cycle. Oncogene 1990; 5:833-837.
87. White AE, Livanos EM, Tlsty TD. Differential disruption of genomic integrity and cell cycle regulation in normal human fibroblasts by the HPV oncoproteins. Genes & Dev 1994; 8:666-677.
88. Zerfaß K, Levy L, Cremonesi C et al. Cell cycle dependent disruption of E2F/p107 complexes by human papillomavirus type 16 E7. J Gen Virol 1995; 76:1815-1820.
89. Hickman ES, Picksley SM, Vousden KH. Cells expressing HPV16 E7 continue cell cycle progression following DNA damage induced p53 activation. Oncogene 1994; 9:2177-2181.
90. Slebos RJC, Lee MH, Plunkett BS et al. p53-dependent G(1) arrest involves pRB-related proteins and is disrupted by the human papillomavirus 16 E7 oncoprotein. Proc Natl Acad Sci USA 1994; 91:5320-5324.
91. Demers GW, Foster SA, Halbert CL et al. Growth arrest by induction of p53 in DNA damaged keratinocytes is bypassed by human papillomavirus 16 E7. Proc Natl Acad Sci USA 1994; 91:4382-4386.
92. Vousden KH, Vojtisek B, Fisher C et al. HPV16 E7 or adenovirus E1A can overcome the growth arrest of cells immortalized with a temperature-sensitive p53. Oncogene 1993; 8:1697-1702.
93. Polyak K, Lee MH, Erdjumentbromage H et al. Cloning of p27(Kip1), a cyclin-dependent kinase inhibitor and a potential mediator of extracellular antimitogenic signals. Cell 1994; 78:59-66.

94. Firpo EJ, Koff A, Solomon MJ et al. Inactivation of a Cdk2 inhibitor during interleukin 2-induced proliferation of human T lymphocytes. Mol and Cell Biol 1994; 14:4889-4901.

95. Zerfass-Thome K, Zwerschke W, Mannhardt B et al. Inactivation of the cdk inhibitor p27KIP1 by the human papillomavirus type 16 E7 oncoprotein. Oncogene 1996; 13:2323-2330.

96. Pan H, Griep AE. Altered cell cycle regulation in the lens of HPV-16 E6 or E7 transgenic mice: implications for tumor suppressor gene function in development. Genes Dev 1994; 8:1285-99.

97. Butz K, Shahabeddin L, Geisen C et al. Functional p53 protein in human papillomavirus-positive cancer cells. Oncogene 1995; 10: 927-936.

98. Peacock JW, Chung S, Bristow RG et al. The p53-mediated G1 check-point is retained in tumorigenic rat embryo fibroblast clones trans-formed by the human papillomavirus type 16 E7 gene and EJ-ras. Mol Cell Biol 1995; 15:1446-54.

99. Spitkovsky D, Aengeneyndt F, Braspenning J et al. p53-independent growth regulation of cervical cancer cells by the papillomavirus E6 oncogene. Oncogene 1996; 13:1027-1035.

100. Boyer SN, Wazer DE, Band V. E7 protein of human papilloma vi-rus-16 induces degradation of retinoblastoma protein through the ubiquitin-proteosome pathway. Cancer Res 1996; 56:4620-4624.

101. Stirdivant SM, Ahern JD, Oliff A et al. Retinoblastoma protein bind-ing properties are dependent on 4 cysteine residues in the protein binding pocket. J Biol Chem 1992; 267:14846-51.

102. Wu EW, Clemens KE, Heck DV et al. The human papillomavirus E7 oncoprotein and the cellular transcription factor E2F bind to sepa-rate sites on the retinoblastoma tumor suppressor protein. J Virol 1993; 67:2402-2407.

103. Dyson N, Guida P, Münger K et al. Homologous sequences in aden-ovirus E1A and human papillomavirus E7 proteins mediate interac-tions with the same set of cellular proteins. J Virol 1992; 66: 6893-6902.

104. Pagano M, Dürst M, Joswig S et al. Binding of the human E2F tran-scription factor to the retinoblastoma protein but not to cyclin A is abolished in HPV-16-immortalized cells. Oncogene 1992; 7:1681-1686.

105. Chellappan S, Kraus V, Kroger B et al. Adenovirus E1A, simian virus 40 tumor antigen, and human papillomavirus E7 protein share the capacity to disrupt the interaction between transcription factor E2F and the retinoblastoma gene product. Proc Natl Acad Sci USA 1992; 89:4549-4553.

106. Arroyo M, Bagchi S, Raychaudhuri P. Association of the human papillomavirus type-16 E7 protein with the S-phase-specific E2F-cyclin-A complex. Mol and Cell Biol 1993; 13:6537-6546.

107. Phelps WC, Bagchi S, Barnes JA et al. Analysis of trans activation by human papillomavirus type 16 E7 and adenovirus 12S E1A suggests a common mechanism. J Virol 1991; 65:6922-6930.
108. Kovesdi I, Reichel R, Nevins JR. Identification of a cellular transcription factor involved in E1A trans-activation. Cell 1986; 45:219-228.
109. Lam EWF, Morris JDH, Davies R et al. HPV16 E7 oncoprotein deregulates B-Myb expression—correlation with targeting of P107/E2F complexes. EMBO J 1994; 13:871-878.
110. Melillo RM, Helin K, Lowy DR et al. Positive and negative regulation of cell proliferation by E2F-1: Influence of protein level and human papillomavirus oncoproteins. Mol and Cell Biol 1994; 14:8241-8249.
111. Lukas J, Müller H, Bartkova J et al. DNA tumor viruses oncoproteins and retinoblastoma gene mutations share the ability to relieve the cell's requirement for cyclin D1 function in G1. J Cell Biol 1994; 125:625-638.
112. Nichols GE, Williams ME, Gaffey MJ et al. Cyclin D1 gene expression in human cervical neoplasia. Modern Pathology 1996; 9:418-425.
113. Cavanaugh AH, Hempel WM, Taylor LJ et al. Activity of RNA polymerase I transcription factor UBF blocked by Rb gene product. Nature 1995; 374:177-180.
114. White RJ, Trouche D, Martin K et al. Repression of RNA polymerase III transcription by the retinoblastoma protein [see comments]. Nature 1996; 382:88-90.
115. Antinore MJ, Birrer MJ, Patel D et al. The human papillomavirus type 16 E7 gene product interacts with and trans-activates the AP1 family of transcription factors. EMBO J 1996; 15:1950-1960.
116. Wong HK, Ziff EB. The human papillomavirus type 16 E7 protein complements adenovirus type 5 E1A amino-terminus-dependent transactivation of adenovirus type 5 early genes and increases ATF and Oct-1 DNA binding activity. J Virol 1996; 70:332-340.
117. Mazzarelli JM, Atkins GB, Geisberg JV et al. The viral oncoproteins Ad5 E1A, HPV16 E7 and SV40 TAg bind a common region of the TBP-associated factor-110. Oncogene 1995; 11:1859-1864.
118. Münger K, Yee C, Phelps W et al. Biochemical and biological differences between the E7 oncoproteins of the high- and low-risk human papillomavirus types are determined by amino-terminal sequences. J Virol 1991; 65:3943-3948.
119. Ibaraki T, Satake M, Kurai N et al. Transacting activities of the E7 genes of several types of human papillomavirus. Virus Genes 1993; 7:187-196.
120. Zajchowski DA, Boeuf H, Kedinger C. E1a inducibility of the adenoviral early E2a promoter is determined by specific combinations of sequence elements. Gene 1987; 58:243-256.
121. Rawls JA, Pusztai R, Green M. Chemical synthesis of human papillomavirus type 16 E7 oncoprotein: autonomous protein domains

for induction of cellular DNA synthesis and for trans activation. J Virol 1990; 64:6121-9.

122. Morosov A, Phelps WC, Raychaudhuri P. Activation of the c-fos gene by the HPV16 oncoproteins depends upon the cAMP-response element at -60. J Biol Chem 1994; 269:18434-18440.

123. Morozov A, Subjeck J, Raychaudhuri P. HPV16 E7 oncoprotein induces expression of a 110 kDa heat shock protein. FEBS Letters 1995; 371:214-218.

124. Nishida M, Miyamoto S, Kato H et al. Transcriptional repression of smooth-muscle α-actin gene associated with human papillomavirus type 16 E7 expression. Molecular Carcinogenesis 1995; 13:157-165.

125. Tommasino M, Adamczewski JP, Carlotti F et al. HPV16 E7 protein associates with the protein kinase p33CDK2 and cyclin A. Oncogene 1993; 8:195-202.

126. Mcintyre MC, Ruesch MN, Laimins LA. Human papillomavirus E7 oncoproteins bind a single form of cyclin E in a complex with cdk2 and p107. Virology 1996; 215:73-82.

127. Shirodkar S, Ewen M, DeCaprio JA et al. The transcription factor E2F interacts with the retinoblastoma product and a p107-cyclin A complex in a cell cycle-regulated manner. Cell 1992; 68:157-166.

128. Xiong Y, Kuppuswamy D, Li Y et al. Alteration of cell cycle kinase complexes in human papillomavirus E6- and E7-expressing fibroblasts precedes neoplastic transformation. J Virol 1996; 70:999-1008.

129. Khleif SN, DeGregori J, Yee CL et al. Inhibition of cyclin D-CDK4/CDK6 activity is associated with an E2F-mediated induction of cyclin kinase inhibitor activity. Proc Natl Acad Sci USA 1996; 93:4350-4.

130. Jewers R, Hildebrandt P, Ludlow J et al. Regions of HPV 16 E7 oncoprotein required for immortalization of human keratinocytes. J Virol 1992; 66:1329-1335.

131. Ohtsubo M, Theodoras AM, Schumacher J et al. Human cyclin E, a nuclear protein essential for the G(1)-to-S phase transition. Mol and Cell Biol 1995; 15:2612-2624.

132. Howes KA, Ransom N, Papermaster DS et al. Apoptosis or retinoblastoma: alternative fates of photoreceptors expressing the HPV-16 E7 gene in the presence or absence of p53. Genes Dev 1994; 8:1300-10.

133. Pan HC, Griep AE. Temporally distinct patterns of p53-dependent and p53-independent apoptosis during mouse lens development. Genes & Dev 1995; 9:2157-2169.

134. Puthenveettil JA, Frederickson SM, Reznikoff CA. Apoptosis in human papillomavirus 16 E7, but not E6-immortalized human urorpithelial cells. Oncogene 1996; 13:1123-1131.

135. Isacson C, Kessis TD, Hedrick L et al. Both cell proliferation and apoptosis increase with lesion grade in cervical neoplasia but do not correlate with human papillomavirus type. Cancer Res 1996; 56:669-74.

136. Lees E. Cyclin dependent kinase regulation. Curr Opin Cell Biol 1995; 7:773-780.

CHAPTER 5

Immunological Aspects of the E6 and E7 Oncogenes: Tools for Diagnosis and Therapeutic Intervention

Ingrid Jochmus and Lutz Gissmann

5.1. INTRODUCTION

The antibody response against structural HPV proteins that was shown to develop during the natural course of HPV infections is likely the consequence of "mucosal immunization" (i.e., uptake and presentation by M cells), or depends on "secondary infection" through small wounds resulting in the exposure of these proteins to antigen presenting cells (APC) such as macrophages or dendritic cells. There is good evidence that antibodies against HPV capsid proteins are relevant in controlling HPV infections.[1]

In contrast, antibodies to early HPV proteins such as E6 and E7 which are constitutively expressed in HPV infected epithelial cells obviously do not influence the outcome of infection. They seem to be a marker of HPV-related malignant tumors that develop as a late consequence of virus persistence. Papillomaviruses are strictly epitheliotropic and apparently no viremic phase is involved at any point during virus replication. Since epithelial cells

Papillomaviruses in Human Cancer: The Role of E6 and E7 Oncoproteins, edited by Massimo Tommasino. © 1997 Landes Bioscience.

are not professional APC, and therefore unable to support the induction of anti-virus immune reactions, exposure of the HPV early proteins to the immune system is probably due to necrosis of tumor cells during development of invasive cancer.

The proteins E6 and E7 are expressed in HPV infected epithelia in the absence of efficient cellular immune reactions, indicating the ability of the virus to escape immune surveillance by the host. Nevertheless, cellular immune mechanisms are most likely involved in the control of papillomavirus infections as it is emphasized by the presence of CD4 positive cells and macrophages in regressing papillomas,[2] and by the high prevalence of clinically manifest HPV infections in immunosuppressed individuals.[3-5] The mechanism(s) triggering an effective cell-mediated immunity against HPV related lesions are barely understood. Activation of the immune system by exposure to viral antigens (following treatment) or by inflammation may be equally important as the genetic composition of the host. Understanding of these processes is of particular importance for the development of a therapeutic vaccine.[6,7]

5.2. HUMORAL IMMUNE RESPONSE AGAINST E6 AND E7

5.2.1. Principles of serological assays

Detection of HPV nonstructural proteins in cells of naturally infected tissue (e.g., in warts) by in situ staining techniques was shown to be extremely difficult. Therefore the E6 and E7 proteins produced during natural infection are inapplicable as antigens for serological assays, which instead depend on the production of viral proteins (or of epitopes derived thereof) by recombinant DNA technology or chemical synthesis. So far, in most instances investigators used either complete proteins produced in various hosts (or by in vitro translation) or chemically synthesized peptides.

5.2.1.1. Western Blotting

The most common assays employed for the detection of HPV-specific antibodies in human sera are Western blotting (WB) or enzyme-linked immunosorbent assay (ELISA). The WB technique

(Fig. 5.1) is a time-consuming procedure and thus is less suitable for analysis of large numbers of specimens. It includes electrophoresis of the protein under denaturing conditions through SDS-poly-acrylamide-gels followed by transfer to a membrane. The membrane is cut into small stripes each of which is incubated with an individual serum usually at dilutions between 1:10 and 1:100. If antibodies are bound to the protein they can be visualized by enzyme-coupled secondary antibodies which are specific for human γ-globulins. The membrane is finally incubated with a suitable substrate for the enzyme which induces a color reaction at the position where the reactive protein was banded. Thus the WB technique permits the identification of a specific antigen-antibody

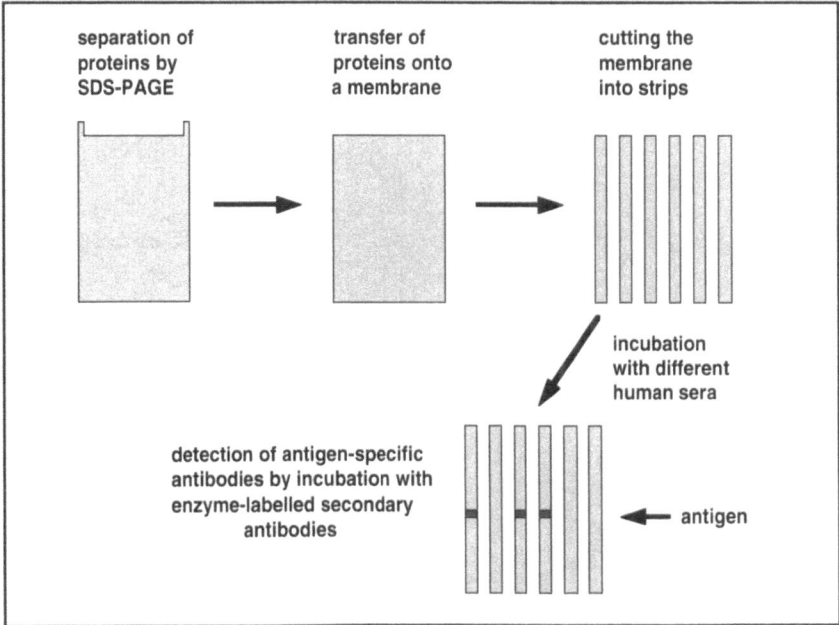

Fig. 5.1. Detection of HPV-specific antibodies by Western blotting. Bacterial fusion proteins are separated by SDS-gel elctrophoresis and transferred to membranes (nitrocellulose or nylon). About 2 mm wide membrane-strips are incubated with individual human sera (diluted between 1:10 to 1:100). For detection of antigen/antibody complexes secondary anti-human immunoglobulin antibodies labeled with horseradish peroxidase or alkaline phosphatase are used. These enzymes catalyze color reactions of appropriate substrates visible as precipitation at the position of the antigen/antibody complex.

reaction even if the protein was not completely purified prior to the electrophoresis. The strength of the reaction (indicated by the intensity of the color reaction) correlates to the amount of reactive antibodies which were present in the sample, but monitoring is only semiquantitative. Since the protein is bound to the membrane in a denatured form, antibodies to sequential (linear) rather than discontinuous (conformational) epitopes are measured (Fig. 5.2).

5.2.1.2. ELISA

Enzyme-linked secondary antibodies are also used in ELISAs (Fig. 5.3) to monitor the antigen-antibody reaction. Since in this case the intensity of the color reaction is measured photometrically, the amount of the antibodies bound to the antigen can within a certain range be quantified in relation to a standard serum. The antigen is coated onto the walls of a microtiter plate either directly

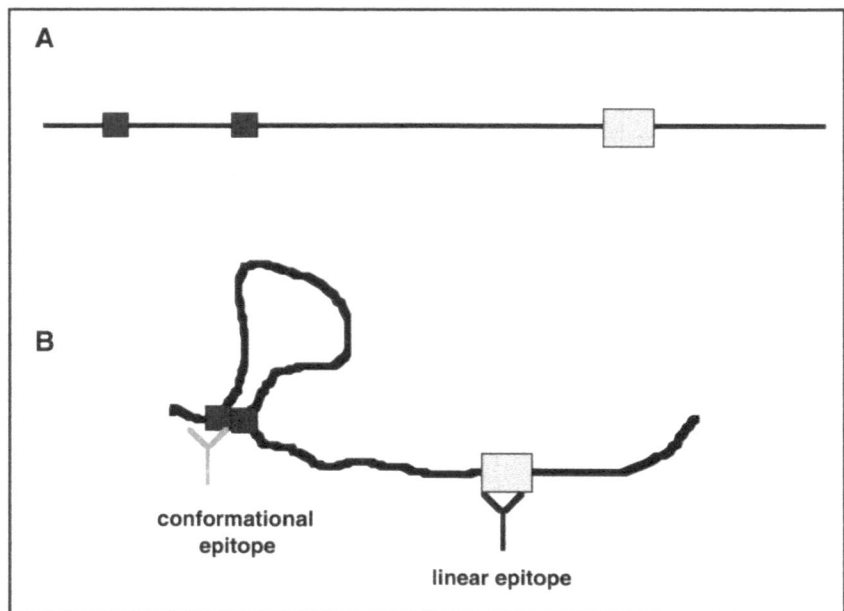

Fig. 5.2. Schematic representation of linear and conformational epitopes. Linear epitopes are defined by the amino acid-sequence of a protein, whereas conformational epitopes depend on the proper folding of the protein. (A) unfolded protein; (B) folded protein.

or in case of complete proteins (see below) through a monoclonal (capture) antibody. The latter variant results in a highly specific binding of the antigen thus in reduction of noise due to unwanted reactions between contaminating (e.g., *E. coli*-derived) proteins and serum antibodies. Also, a possible denaturation as a consequence of the direct binding can most likely be avoided. On the other hand capture ELISA may suffer from loss of some reactivity as eventually the monoclonal antibody may cover certain epitopes of the protein. By ELISA many samples can be tested simultaneously and the processing is partly automated. Hence this assay is well suited for large-scale screening of serum samples. Either whole proteins or synthetic peptides derived thereof have been used as antigens. Synthetic peptides represent well defined reagents and peptide-based ELISAs eliminate the risk of false positive results due to cross reactions with contaminating nonviral products such as bacterial cell components (in case *E. coli* made fusion proteins are applied). Synthetic peptides proved to be very handy to identify seroreactive epitopes, and in a large number of studies informative data were accrued when peptide-specific ELISAs were applied. On the other hand, antibodies directed against conformational epitopes are very likely not detected by synthetic peptides. Therefore, the application of complete proteins does, as expected, reduce the number of

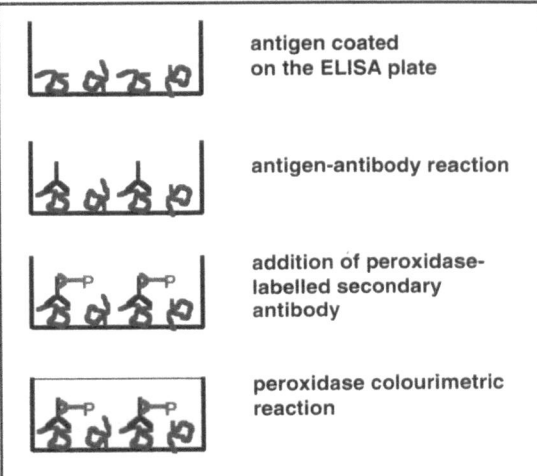

antigen coated
on the ELISA plate

antigen-antibody reaction

addition of peroxidase-
labelled secondary
antibody

peroxidase colourimetric
reaction

Fig. 5.3. Detection of HPV-specific antibodies by ELISA. Complete proteins or peptides representing B cell epitopes coated to microtiter wells are used as antigens. Detection of antigen/antibody complexes is similar to Western blot analysis. However, in contrast to this technique in ELISA assays the antigens need to be purified and the amount of antibodies can be quantified photometrically.

false negative samples (increase sensitivity). In addition, in recent studies it was shown that ELISAs using complete proteins are more specific (reduce number of false positives) as compared to peptide-based ELISAs.[8,9] The reason for this observation may be that the linear epitopes represented by the peptides are in some instances more broadly cross reacting with antibodies to the proteins of other HPV types.

5.2.1.3. RIPA

Recently, radio-immunoprecipitation (RIPA) (Fig. 5.4) was introduced for the detection of HPV E6 or E7-specific antibodies.[10] This method takes advantage of the fact that antigen-antibody complexes can easily be precipitated when they react with Protein A of *Staphylococcus aureus* or with a secondary antibody

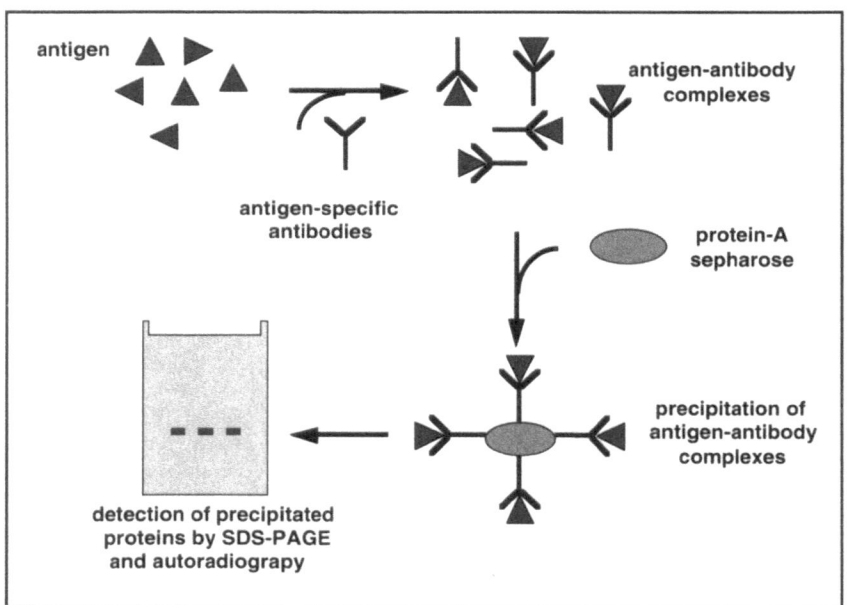

Fig. 5.4. Detection of antibodies against conformational epitopes of HPV antigens by radio-immunoprecipitation assay. Radiolabeled proteins are produced by in vitro transcription and translation and incubated with human sera (dilution 1:100). Protein-A sepharose dependent precipitation of the antigen occurs only in the presence of antigen-specific antibodies. After disruption of the precipitates by boiling the antigens are separated by SDS-gel electrophoresis and subsequently detected by autoradiography.

which are coupled to a carrier such as Sepharose. The precipitate is subsequently dissociated and separated by SDS-polyacrylamide-gel electrophoresis. When radioactively labeled viral protein (in case of HPV produced by an in vitro transcription/translation protocol) is used it can be monitored with high sensitivity by autoradiography. Very likely antibodies against linear and conformational epitopes are detected by RIPA. The amount of precipitated radioactive protein can be quantitatively measured. However, this assay is too time-consuming to be useful for routine testing. A modification of the RIPA was described by Nindl et al who used unlabeled HPV16 E7 protein expressed in vaccinia for precipitation. The HPV protein was detected by Western blot analysis with the aid of an E7-specific monoclonal antibody.[8]

5.2.1.4. Use of serology to diagnose HPV infection

The power of a diagnostic assay is defined by the number of positive reactions among infected individuals compared to a control population. Ideally, the assay will score positive for all the infected individuals (100% sensitivity) but in none of the controls (100% specificity). In viral diagnostics valid assays were developed in those cases where infection can easily be assessed by specific symptoms such as rubella. However, serology of HPV infection is complicated by the following issues and it is sometimes impossible to assess whether false (positive or negative) results are due to technical flaws or the consequence of the biologic properties of the papillomaviruses: (i) HPV infections are in many instances subclinical and, therefore, control populations are difficult to define. (ii) Papillomavirus infection occurs at an immunologically privileged site (i.e., the epithelium), and it is unclear whether during the natural history of an infection an immune response is induced in all instances. (iii) There is some cross-reactivity between the viral proteins of different HPV types and multiple papillomavirus infections are common.

For these reasons the antibody profile during the natural course of papillomavirus infections is only poorly defined. There is good evidence, however, that the humoral immune response to the early proteins E6 or E7 of the well characterized "high risk" HPV types 16 and 18 is only developing as late consequence of persistent infection and seems to concur with the appearance of invasive cervical cancer. The following summary will be confined to this topic.

5.2.2. Antibodies to HPV E6 or E7 proteins

5.2.2.1. Correlation of antibodies against E6 and E7 with cervical cancer

The correlation of HPV16 E7-specific antibodies with cervical cancer was first published in 1989[11] and is now well established in the literature.[12-15] A similar association was subsequently found for E6-specific antibodies.[10,16-18] As of today antibodies to the proteins E6 or E7 of HPV16 (or HPV18) can be considered as a marker for cervical cancer and possibly for recurrent disease after treatment of the tumor (P. Herbrink, personal communication).[19,20] It remains to be investigated, however, whether serology is of use for clinical diagnosis. In contrast, no function in control of the disease has yet been assigned to the humoral immune response.

In a study by Jochmus-Kudielka et al sera were analyzed by Western blot analysis using HPV16 E7 fusion proteins expressed in *E. coli*.[11] Antibodies to this protein were found 14 times more frequently in cancer patients than in age-matched controls ($p < 10-5$). Suchankova et al found HPV16 E7-specific antibodies in 12/34 cervical cancer patients and in 6/107 controls and noted a good correlation of the data obtained by WB and peptide-ELISA.[21] Köchel et al detected antibodies to E7 of HPV16 and/or to E7 of HPV18 in 29/46 (63%) of cervical cancer patients but only in 4/41 (10%) of healthy individuals.[22] Mandelson et al described an association with cervical cancer ($RR = 3.8$; $CI 95\% = 1.7$-8.5) for strong reacting sera.[23] Kanda et al and Onda et al reported an association of antibodies to HPV16 E7 with cervical cancer sera independent of antibodies to the HPV16 E4 protein.[24,25] Association with cervical cancer of antibodies to the E7 of HPV16 was also found by Paez et al[26] (18% of patients, n = 56 vs. 5% of controls, n = 200; $p < 0.01$).

The first study describing the use of an E7-specific ELISA was published by Mann et al.[27] These investigators reported an increased antibody prevalence against the complete (chemically synthesized) E7 protein in cervical cancer patients (25%; n = 186) as compared to age-matched controls (6%; n = 172). After adjustment for other risk factors the odds ratio of cervical cancer associated with IgG to E7 was 5.3 (95% CI = 2.4-11.6) and with IgA to E7 2.7 (95% CI = 1.3-5.3). HPV16 E7 (and E6)-specific peptides were also used by Müller et al,[10] who found 37% (10%) of patients with HPV16 DNA positive tumors (n = 39) and 9% (1%) of controls (n = 39) positive by ELISA (p < 10-5 in both cases). Bleul et al found antibodies to an HPV18 E7-peptide in sera of 10/116 cervical cancer patients but in none of 116 controls (p < 0.001).[28] Type-specificity of the reaction was suggested by the analysis of HPV18 DNA in the patients' biopsies as well as by testing of the sera with a peptide derived from the homologous HPV16 E7 region. Type-specific reaction to an HPV18 E7-derived peptide was also reported by Krchnak et al who analyzed pooled sera originating from patients with HPV18 and HPV16 positive cervical cancer or with genital warts.[29] Iha et al found that HPV16 E7- and/or HPV18 E7-specific antibodies were significantly related to cancer risk (RR 1.9; 95% CI 1.2-2.3; 219 cases, 387 controls).[30] The immune response to HPV proteins was shown to be the only independent risk factor in comparison to antibodies to other sexually transmitted agents (e.g., herpes simplex virus 2 and *Chlamydia trachomatis*). In a population-based case-control study, Dillner et al tested sera from 94 incident cases of cervical cancer and from 188 age- and sex-matched controls against a panel of peptide antigens derived from HPV types 6, 11, 16 and 18.[31] Significantly increased risks (RR) were found for IgG type antibodies to HPV16- or HPV18-derived peptides from the open reading frames E7 (RR = 3.8 and 2.7) and for IgA antibodies to HPV peptides from HPV16 E6 (RR = 2.7). Sasagawa et al used HPV16 E6 and E7 fusion proteins in ELISA and found antibodies to E6 (E7) in 27% (33%) of cervical cancer patients (n = 30) and in none of sera from healthy women.[32]

The HPV16 E6 and E7 protein produced by in vitro transcription-translation were first used by Müller et al and Viscidi et al who analyzed by radio-immunoprecipitation (TT-RIPA) sera obtained from two case control studies in Colombia and Spain.[10,16] Antibodies to the E6 and E7 protein were observed among 56% and 43% (72% to either protein) of invasive cervical cancer cases and among 1.7% and 4.1% (5.8%) of controls.[16] Comparative analysis of a set of 137 cervical cancer sera by RIPA and by peptide ELISA revealed similar sensitivity to the HPV16 E7 protein (31% versus 39%) but a higher sensitivity to HPV16 E6 by RIPA (47% versus 17%).[18] In a case control study of cervical cancer in Brazil antibodies to the HPV16 E6 (E7) protein were found in 54% (30%) of 194 cases and 6% (5%) of 217 controls.[15] Reactivity to the E6 protein was shown to be associated with stage of disease (30% stage I vs. 63% stages II-IV; p < 0.005). Increased sensitivity of E7-specific antibodies across tumor stages was found in two other studies.[19,33]

5.2.2.2. Why do not all cervical cancer patients have antibodies against HPVE6 and/or E7?

Surprisingly, antibodies to both E6 and E7 are only found in up to 50 to 70% of sera obtained from cervical cancer patients with proven HPV16- or HPV18-positive cancer biopsies. This cannot be explained by naturally occurring variants of HPV16 or HPV18 E7 as no correlation was found between seroreactivity and the presence of such variants identified within the tumor of the individual patients.[34,35] Further, it is unlikely that the negative results are due to experimental limitations of the serological assays employed since a subset of sera remains negative even when complete viral proteins in a native form rather than synthetic peptides or denatured proteins are used for antibody detection.[15] Finally, there is a fraction of negative patients even with advanced stages of the disease, although recent studies suggest that the humoral immune response depends upon the tumor size as the number of positive sera and antibody titers rise across tumor stage and drop after treatment of the disease.[19,20,33]

From these data it can be concluded that the failure of certain cervical cancer patients to react with HPV early proteins is very likely not due to technical limitations of the assays, genetic variability of the virus or to a short exposure of the viral proteins to the immune system. It rather appears that certain patients actually fail to develop HPV-specific antibodies. Recently, positive or negative associations between particular HLA class II haplotypes and the occurrence of cervical cancer or the presence of certain HPV types was reported.[36-41] It remains to be elucidated whether there is a correlation between HLA polymorphism and HPV seroreactivity as well. By the detection of viral antibodies directed against different HPV proteins (e.g., E4, L1) as well as of HPV DNA within oral or genital swabs it was suggested that infection with HPV16 may occur perinatally or during early infancy.[11,42,43] In case of HPV6 or HPV11 it was reported that the early infection eventually may lead to the development of laryngeal papillomatosis,[43a] but the clinical consequence of infection with HPV16 remains unclear. It is speculated, however, that in most instances papillomavirus infections early in life stay inapparent and under certain circumstances may render this particular individual tolerant. Obviously, this hypothesis cannot be verified in human beings, but experiments using HPV16 transformed cells in mice showed that tolerance against the HPV16 E7 protein can actually be induced.[44]

5.3. CELLULAR IMMUNE RESPONSES AGAINST E6 AND E7

5.3.1. Methods to measure cellular immune responses

In T lymphocyte proliferation assays using either spleen or lymph node cells of immunized mice, or human peripheral blood lymphocytes (PBL) as responder cells, the CD4 positive T cell reactivity is mainly monitored. The lymphocytes are cultured in the presence of antigens. After three to five days, the antigen induced T cell proliferation can be measured by ^{3}H-thymidine incorporation (Fig. 5.5). If the frequency of reactive T cells is too low to obtain significant proliferation upon direct stimulation, the T cells can be restimulated with syngeneic irradiated antigen-loaded stimulator cells after seven days and incubated for another three to five days prior to the addition of ³H-thymidine. By this method epitopes within the E6 and E7 open reading frames have been identified that are recognized by CD4 positive T cells of mouse or human origin in combination with different MHC class II molecules.

Delayed type hypersensitivity is mediated by CD4 positive T lymphocytes and can be monitored in animal models after challenge with antigen by measuring ear swelling reactions. In humans a skin test comparable to the tuberculin test, in which small amounts of the antigen are delivered underneath the skin, would be the corresponding method.[45]

Specific reactivity of cytotoxic T lymphocytes (CTL) can be measured (i) either in vitro in ⁵¹Cr release assays after one or several restimulations in culture, or (ii) in vivo in tumor protection

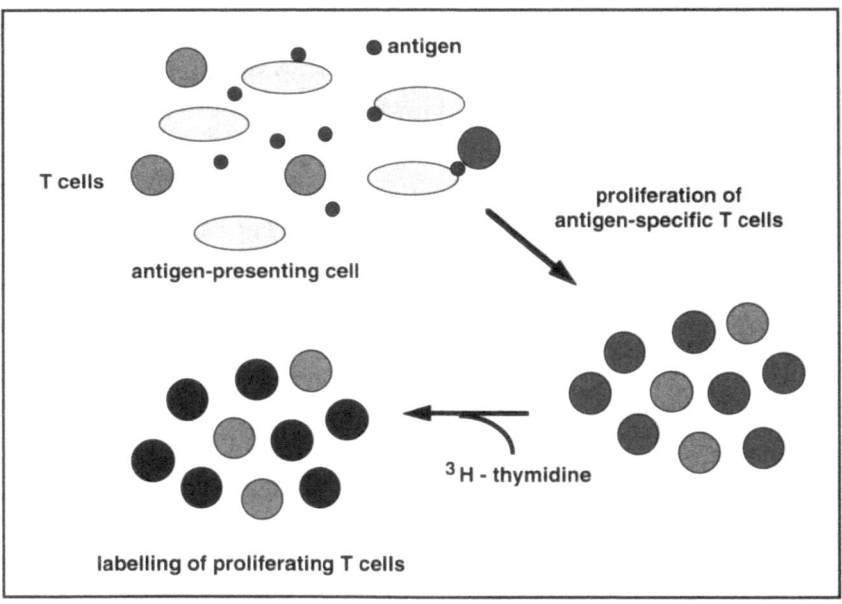

Fig. 5.5. T cell proliferation assay. In proliferation assays proteins or peptides are incubated in microtiter wells with PBL containing T and B lymphocytes as well as antigen presenting cells (APC). The antigens are taken up by APC, processed and presented to (mainly CD4-positive) T cells which start to proliferate, if they have an antigen-specific receptor (black cells). Proliferation is measured by incorporation of tritiated thymidine.

experiments. The CTL can be phenotypically characterized by FACS analysis and in CD8 depletion experiments. (i) During restimulation with syngeneic irradiated stimulator cells presenting a specific antigen, CTL recognizing this antigen are activated and induced to proliferate. The restimulations are performed weekly. In ^{51}Cr release assays, CTL effector cells are incubated with ^{51}Cr labeled antigen-expressing or antigen-loaded target cells for four to six hours. The ^{51}Cr released by the target cells during this incubation period corresponds to the specific lytic activity of the effector cells (Fig. 5.6). (ii) In tumor protection experiments, animals are immunized, and stimulation of specific CTL is demonstrated by inhibition of tumor development after inoculation of tumorigenic cells expressing the antigen that was used for immunization.

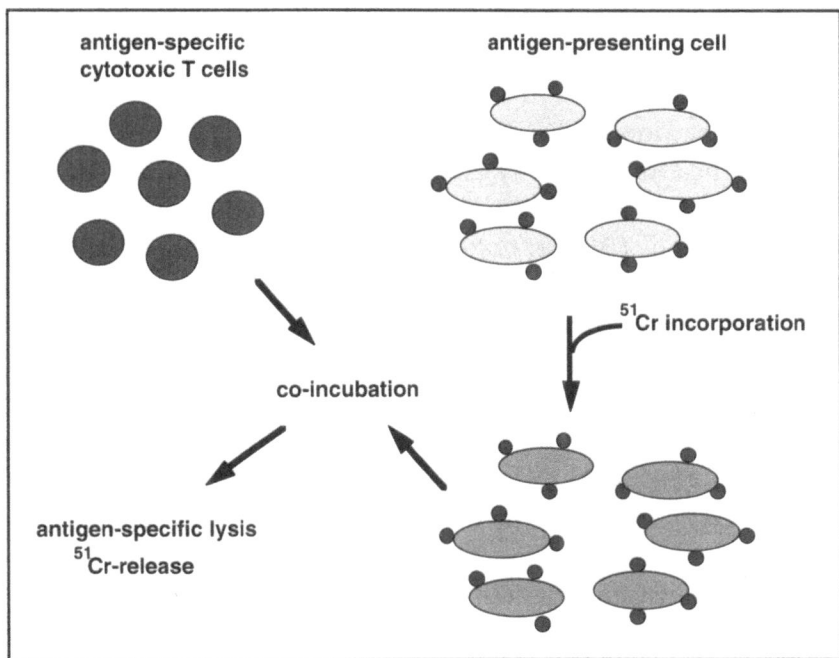

Fig. 5.6. Cytotoxicity assay. In cytotoxicity assays cells presenting either externally loaded peptides (CTL epitopes) or peptides from intracellularly expressed proteins are used as target cells. After labeling with ^{51}Cr the target cells are incubated in microtiter wells with cytotoxic T (effector) cells in different effector/target ratios. Lysis of the target cells is quantified by measuring in a β-counter the release of ^{51}Cr into the medium.

Therapeutic effectiveness of the CTL becomes obvious when already established tumors are eliminated.

5.3.2. T lymphocyte proliferation

5.3.2.1. Mouse T cell proliferation

In an early study of Commerford et al CBA mice (H2k) were immunized with HPV16 E7 expressed as fusion protein in bacteria.[46] Lymphoproliferative assays were performed with lymph node cells and overlapping 20-mer peptides as antigens, that covered the whole E7 protein. The authors found four T cell epitopes localized at amino acids (amino acid) 20-29, 45-54, 60-79 and 85-94. In a similar approach, but with different mouse strains five immunodominant T cell stimulatory regions within E7 were defined.[47] Although the mouse strains showed different response profiles to individual peptides, T lymphocytes of most of them recognized peptides derived from these immunodominant regions, four of which (amino acid 17-32, 42-59, 62-77, 87-98) correspond to the epitopes identified by Commerford et al.[46] The E7 epitope DRAHYNI (amino acid 48-54) had been described by Tindle et al to be a "public" T helper epitope.[48] In immunization experiments this epitope was recognized in combination with five MHC class II I-A and I-E alleles, and was able to provide help for the production of antibodies to different B cell epitopes linked to it. Mouse T cell proliferative responses against the epitope were also induced after administration as chimeric hepatitis B core antigen particle or as conjugate of an immunostimulatory carrier (ISCAR).[49,50] The T lymphocytes produced the cytokines IL2 and IL4, indicating the activation of Th1 as well as Th2 helper cells.

Immunization of C57BL/6 mice (H2b) with E6 recombinant vaccinia viruses induced specific T cell proliferation against the E6 derived peptides amino acid 41-50, 91-100 and 146-151.[51] However, attempts to find proliferative responses in DBA mice (H2d) were unsuccessful in this study. This suggests that compared to E7 the immunostimulatory functions of this protein are more restricted.

5.3.2.2. Human T cell proliferation

Almost all regions on HPV16 E6 and E7 that are recognized by mouse CD4 T cells were also immunodominant epitopes for T cells of human origin. While in an early study in which HPV16 and 18 E6 fusion proteins were used as antigens, no significant difference was observed between the lymphoproliferative responses of CIN patients and healthy controls,[52] later investigations found cell mediated immune reactions against the HPV16 E7 derived peptide amino acid 72-97 significantly more frequent in patients attending a colposcopy clinic for evaluation of abnormal cervical cytology and infected with the HPV types 16, 31 or 33 (45.5%) compared to patients infected with other HPV types (6.4%) or controls (7.7%).[53] Other peptides were also stimulatory (amino acid 17-37, 37-54 and 62-80), but the type-specificity of these T cell reactions was statistically not significant. Peptides amino acid 17-37 and 62-80 are overlapping with two of three regions of the HPV16 E7 protein (amino acid 5-18, 17-38 and 69-86) that were previously described by Altmann et al to induce proliferation of T cells of two donors who did not have a known history of HPV infection, but were seropositive against HPV16 E7 in Western blot experiments.[54] Some of the established CD4 positive T cell clones recognizing epitopes of the C-terminal region (amino acid 69-82 and 73-86) showed cytotoxic activity towards peptide loaded target cells. In a recent paper de Gruijl et al reported HPV16 E7 specific memory T cells to be present in patients with cervical intraepithelial neoplasia (CIN) with persistent, cleared and fluctuating HPV16 infections (57.7%), but not in HPV negative patients, using a GST-E7 fusion protein as antigen.[55] The highest proliferation was observed in patients with persistent infection and stable or progressive cervical lesions (88.9%), suggesting that the T helper memory depends on continuous E7 expression, and that the proliferative responses are not predictive for the outcome of the disease. Since T cells of patients infected with other types than HPV16 reacted less frequently in the proliferation assays than T cells of HPV16 positive patients, the reactivity observed was interpreted as HPV type specific. These data are consistent with the findings of Kadish et al.[53] Proliferation against HPV16 E6 could be achieved by stimulating PBL of one healthy asymptomatic donor with the HPV16 E6 derived peptide amino acid 42-57 presented on HLA-DR transfected mouse L cells.[56] In all cases[52-56] it cannot be definitively decided whether the observed lymphoproliferation was actually generated by restimulation of memory T cells or by in vitro priming of naive T cells. Future studies investigating larger and better characterized patient and control groups will clarify this issue, and answer the question if HPV infection induces memory CD4 T cell responses, and if T cell proliferation assays can be useful tools for the diagnosis of HPV associated diseases.

5.3.3. Delayed type hypersensitivity

Delayed type hypersensitivity (DTH) against HPV16 E7 was observed in a mouse model in which HPV16 transformed mouse keratinocytes were grafted on syngeneic mice permitting reformation of the differentiated epithelium.[57] CD4 positive DTH T cells were involved in the early rejection of the grafted cells after challenge with a recombinant vaccinia virus expressing the HPV16 E7 gene as determined by measuring ear swelling reactions and by CD4 depletion experiments. Similar results were obtained for HPV16 E6.[58] Interestingly, grafting with a low number of HPV16 expressing cells induced immunological unresponsiveness against E7, although an epithelium reformed.[44] The mice remained unresponsive to three challenges with E7 over a period of 31 days. Furthermore, the unresponsiveness appeared to be correlated with increased graft survival (24 days). These findings are especially important with respect to the hypothesis that constitutive expression of low levels of HPV antigens after infection induces T cell tolerance and that in this way the virus evades immune eradication.

5.3.4. Cytotoxicity

5.3.4.1. Rodent CTL

Mouse CTL against HPV16 E6 and E7 were induced in vitro and in vivo using either peptides, the whole protein, protein expressing cells,[65-67] or recombinant vaccinia viruses as antigens.[59-63,64,68,69]

In the approach of Stauss et al four E6 and four E7 derived 10-mer peptides were found to strongly bind to the H2-Kb molecule.[59] In the peptide set employed the peptides overlapped by only five amino acids indicating that not all possible peptides were represented. Five of the eight peptides (E6 amino acid 36-45, 41-50 and 81-90; E7 amino acid 21-30 and 71-80) induced a CTL response in vitro. CTL against E6 amino acid 41-50 lysed target cells loaded with a 9-mer lacking one N-terminal amino-acid, but still containing the K[b] anchor residue at the C-terminus, approximately 20-fold less efficiently than targets pulsed with the immunizing 10-mer. This indicates that peptides chosen for vaccination must correspond in length to naturally processed peptides. Peptides that did not bind to H2-Kb were also unable to stimulate CTL. Using a complete library of 240 overlapping synthetic 9-mer peptides with an eight amino-acid overlap covering the whole HPV16 E6 and E7 proteins, Feltkamp et al identified 43 peptides that bind to H2-Kb and/or Db molecules.[60] Five peptides already bind at a very low concentration (0.5 µM) and were therefore regarded as suitable vaccine candidates (H2-Kb: E6 amino acid 72-80; H2-Db: E6 amino acid 109-117, E7 amino acid 49-57 and 57-65; H2-Kb and Db: E6 amino acid 37-45). Immunization of C57Bl/6 mice with the peptide E7 amino acid 44-62, which contains the T helper epitope amino acid 48-54[48] and the potential cytotoxic T cell epitope amino acid 49-57, prevented the development of tumors after inoculation of the HPV16 transformed syngeneic cell line C3. In ongoing experiments the peptide amino acid 49-57 turned out to be sufficient for tumor protection. Effective CTL responses could be achieved by either subcutaneous inoculation of the peptide with incomplete Freund's adjuvant, or intravenous infusion of peptide pulsed syngeneic spleen derived dendritic cells.[60,63] E7 amino acid 49-57 also sensitized syngeneic target cells for lysis and induced specific CTL in vitro. These CTL recognized a peptide fraction eluted from H2-b molecules of EL4 cells infected with E7 recombinant vaccinia viruses which demonstrates that E7 amino acid

49-57 is a natural product of intracellular protein processing.[62] After adoptive transfer, CTL stimulated in vitro with E7 amino acid 49-57 were able to eradicate established C3 tumors in B6 nude mice.[61] However, CTL raised against C3 cells did not recognize E7 amino acid 49-57, but another not yet defined CTL epitope, which suggests that E7 amino acid 49-57 is only subdominant on C3 cells. The peptide E7 amino acid 21-28 which binds to H2-Kb and activated specific CTL in vitro, is apparently not presented on cells expressing the E7 protein.[62]

Immunization of H2-d mice (DBA/2 and BALB/c) with either HPV16 E7 recombinant vaccinia viruses or with the E7 protein in the adjuvant MF59 induced an E7 specific cytotoxic CD8 positive T cell response.[64] T cells obtained in both approaches showed the same specificity for E7 recognizing epitopes of the C-terminal half of the protein and the same peptide fractions eluted from E7 recombinant vaccinia virus infected P815 cells. Gao et al stimulated HPV16 E6 specific T lymphocytes in DBA/2 (H2-d) and C57Bl/6 (H2-b) mice using E6 recombinant vaccinia viruses as antigen.[51] They established E6 specific cytotoxic T cell lines and used them to identify an H2-b restricted cytotoxic T cell epitope (amino acid 131-140) from a set of overlapping 10-mer peptides covering the E6 sequence. This epitope was further characterized.[69] The minimal immunogenic sequence amino acid 130-137 is presented by H2-Kb molecules as determined with either H2-Kb or Db transfected P815 cells (H2-d). CTL against amino acid 130-137 recognized four adjacent HPLC fractions prepared from HPV16 E6 expressing RMA cells (H2-b). Subfractionation of the peptides revealed the presence of at least two distinct activities, of which only one corresponded to the epitope at amino acid 130-137. As HPV16 E7[62] the E6 protein contains another peptide stimulating cytotoxic T lymphocytes (amino acid 43-50) which binds even stronger to H2-Kb than amino acid 130-137, however, in the context of the whole E6 protein turned out to be not relevant.

HPV16 E6 or E7 specific cytotoxic T cells were shown to be effective in tumor rejection in vivo. Inoculation of rats with vaccinia virus recombinants expressing HPV16 E6 or E7 retards or prevents the development of tumors in animals which have been challenged with primary rat cells transformed with HPV16 and activated *ras*.[68] Similarly, immunization of mice with syngeneic nontumorigenic fibroblasts transfected with the HPV16 E6 or E7 gene confers protection against transplanted HPV16 E6 or E7 positive tumor cells.[65,66] Even the tumor cells themselves induce anti-tumor immunity after injection when they express both, HPV16 E7 and the murine B7 antigen.[67] This protection is mediated by CD8 positive cytotoxic T lymphocytes, since treatment of the mice with anti-CD8 antibodies prevents tumor rejection.[66]

5.3.4.2. Human CTL

There are recent studies indicating that HPV E6 and/or E7 specific CTL responses also exist in humans. Binding properties of nonamer peptides derived from the HPV16 E6 and E7 proteins to the HLA class I alleles A*0101, A*0201, A*0301, A*1101, and A*2401 have been carefully analyzed by Kast et al[70] using purified class I molecules and radiolabeled synthetic peptides as probes. In a later study, three of the HLA-A*0201 binding E7 derived peptides (amino acid 11-20, 82-90 and 86-93) were shown to induce human CTL from healthy donors in vitro, and HLA-A*0201 restricted mouse CTL after immunization of transgenic mice expressing a HLA-A2Kb hybrid molecule.[71] The human CTL were able to cross-react with the HLA-A*0201 matched cervical carcinoma cell line CaSki, however, since E7 negative CaSki cells are not available as control, there is no definite proof of the in vivo relevance of these peptides. In fact, in our investigations E7 86-93 specific CTL that lysed CaSki cells, were unable to recognize other E7 expressing and HLA-A*0201 positive cells.[71a] By testing the reactivity of PBL of HPV16 positive CIN (n = 11) and cervical cancer patients (n = 11), as well as healthy donors (n = 10) in short term stimulation experiments using the peptides E7 amino acid 11-20 and 86-93 as antigens, only two CIN and two cervical cancer patients, but none of the healthy donors had memory CTL against the peptide amino acid 11-20.[72] Reactivity against amino acid 86-93 could not be detected in any of the donors, underlining the suggestion that this peptide does not play a role as CTL epitope in vivo. A naturally processed HLA-*A0201 restricted HPV16 E6 derived peptide was identified by elution of peptide/MHC class I complexes of HLA *A0201 positive cells infected with E6 recombinant vaccinia viruses, but so far a CTL response against this peptide could not be demonstrated.[73] Tarpey et al induced HPV11 E7 specific CTL using low density (dendritic) cells from PBL of healthy donors either loaded with synthetic peptides or presenting peptides after infection with HPV11 E7 recombinant vaccinia viruses.[74] CTL stimulated with endogenously processed HPV11 E7 recognized the peptide amino acid 4-12 which in reciprocal experiments was able to induce CTL that lysed HPV11 E7 expressing cells. Therefore, although the reactivity of the CTL was only weak, it can be assumed that this peptide represents a relevant cytotoxic T cell epitope.

5.4. HPV E6 AND E7 AS POTENTIAL THERAPEUTIC VACCINES

"High risk" HPV types are involved in the development of anogenital cancer, especially cancer of the cervix and its precursor lesions (cervical intraepithelial neoplasia, CIN).[75] The viral transforming proteins E6 and E7 of "high risk" papillomaviruses can be regarded as tumor antigens and therefore are important targets for an active specific immunotherapy. The following three points are probably the main reasons why, in spite of the expression of these viral antigens, HPV infected cells are not eliminated by the immune system:

The viruses exclusively persist in epithelial cells, which do not express costimulatory signals like CD80 or CD54 that are necessary for proper T cell stimulation.

MHC class I molecules are often downregulated on cells of CIN and cervical cancer rendering them less recognizable by T cells.[76]

Virus infection is usually followed by a long period of latent persistence, during which the viral proteins are expressed at very low levels. This together with inadequate antigen presentation by epithelial cells could induce T cell anergy.

Several animal experiments have demonstrated that these problems can be overcome by effective therapeutic vaccination.

Targeting of the E7 protein into endosomal and lysosomal compartments by linking it to the transmembrane and cytoplasmic region of the lysosomal-associated membrane protein LAMP-1 resulted in MHC class II processing and presentation, and a significantly increased stimulation of MHC class II restricted CD4 positive T cells and thereby in addition the induction of E7 specific CTL in C57BL/6 mice.[77] Antigen presenting cells (APC) transfected with the chimeric E7/LAMP-1 gene were also better in vitro stimulator cells for T lymphocytes primed in vivo with an E7 peptide containing a T helper epitope, than APC expressing the E7 protein alone. Immunization of B57Bl/6 mice with recombinant E7/LAMP-1 vaccinia viruses conferred protection against challenge with E7 expressing epithelial tumor cells, and the immune response induced eliminated already established tumors.[78]

As mentioned above, E6 or E7 recombinant vaccinia viruses as well as E6 or E7 positive epithelial tumor cells transfected with the murine B7 gene were able to induce an effective CTL response in mice,[67,68] suggesting that the same strategies could be successfully applied to humans. CD80 transfected CaSki cells have already been shown to significantly better induce allogeneic and also E7 specific T cell responses in vitro than the nontransfected parental cells, and addition of interleukin 7 enhanced this reactivity (Quiao and Kaufmann, unpublished data). In a clinical trial in the UK eight HPV16 positive late state cervical carcinoma patients were treated with HPV16 and 18 E6 and E7 recombinant vaccinia viruses.[79,80] The E6 and E7 open reading frames were expressed as E6/E7 fusion proteins and contained mutations in order to remove their transforming capacity.[79] In all of the patients anti-vaccinia antibodies were induced, three of them developed antibodies against HPV18 E7, and one of three evaluable patients showed HPV18 specific CTL.[80] Although the immune responses measured do not allow conclusions about the clinical effectiveness of the vaccine, they justify further investigation of this therapeutic approach.

Alternatively, synthetic peptides corresponding to potential cytotoxic T cell epitopes characterized in vitro could be used as possible vaccines. Two of the HLA-*A0201 restricted HPV16 E7 derived peptides described by Ressing et al[71] were employed as antigens in combination with a defined T helper epitope and adjuvants to therapy HLA-*A0201 positive cervical cancer patients. However, peripheral memory CTL could not be detected after immunization,[81] although peptide vaccination was very effective when tested in mice.[60,61] The use of peptide loaded autologous dendritic cells or polypeptides consisting of several CTL epitopes might be promising variants of this approach.[63,82]

Another future possibility for the delivery of the E6 or E7 antigen could be the use of recombinant live bacterial (Salmonella, Streptococcus, etc.) or attenuated viral vectors other than vaccinia (adenovirus, poliovirus, etc.). For example, the HPV16 E7 protein expressed as a fusion protein on the surface of *Streptococcus gordonii*, a nonpathogenic commensal organism of the oral cavity, induced production of specific anti-E7 antibodies (IgG) in mice after subcutaneous inoculation with Freund's adjuvant or stable colonization of the bacteria by a single intranasal-loral inoculation.[83,84] Cellular immune responses have not yet been investigated.

Taken together, the results from animal experiments and the preliminary data obtained from clinical trials with cervical cancer patients indicate that therapy using the HPV E6 or E7 antigens as therapeutic vaccines could be a successful means for the treatment of patients with cancerous or precancerous HPV associated lesions.

REFERENCES

1. Kirnbauer R. Papillomavirus-like particles for serology and vaccine development. Intervirology 1996; 39:54-61.
2. Stanley MA, Coleman N, Chambers M. The host response to lesions induced by human papillomavirus. In: Mindel A, ed. Genital Warts and Human Papillomavirus Infection. London: Edward Arnold, 1994:21-44.
3. Benton C, Shahidulah H, Hunter JAA. Human papillomavirus in the imunosuppressed. Papillomavirus Rep 1992; 3:23-26.
4. Fisher SG, Gissmann L. Convergent infections: human papillomavirus and human immunodeficiency virus. In: Advanced Technologies in

Research, Diagnosis and Treatment of AIDS and in Oncology. Antibiot Chemother, Basel: Karger, 1994; 46:134-149.

5. Stockfleth E, Gissmann L. Human papillomavirus and immunosuppression. In: Shiokawa Y, Kitamura T, eds. Global Challenge of AIDS. 1995:113-122.

6. Altmann A, Gissmann L, Jochmus I. Towards HPV vaccination. In: Minson A, Neil J, McCrae M, eds. Viruses and Cancer. Cambridge University Press, 1994:71-80.

7. Campo MS. Vaccination against papillomavirus in cattle. In: zur Hausen H, ed. Human Pathogenic Papillomaviruses. Current Topics Microbiol Immunol, Berlin: Springer Verlag, 1994:255-266.

8. Nindl I, Gissmann L, Fisher SG et al. The E7 protein of the human papillomavirus (HPV) type 16 expressed by recombinant vaccinia virus can be used for the detection of antibodies in sera from cervical cancer patients. J Virol Methods 1996; 62:81-85.

9. Meschede et al. (submitted).

10. Müller M, Viscidi RP, Sun Y et al. Antibodies to HPV16 E6 and E7 proteins as markers for HPV16-associated invasive cervical cancer. Virology 1992; 187:508-514.

11. Jochmus-Kudielka I, Schneider A et al. Antibodies against the human papillomavirus type 16 early proteins in human sera: Correlation of anti-E7 reactivity and cervical cancer. J Natl Cancer Inst 1989; 81:1698-1704.

12. Viscidi R, Shah KV. Immune response to infections with human papillomaviruses. In: Quinn TC, Gallin JI, Fauci AS, eds. Advances in Host Defense Mechanisms. Vol. 8. New York: Raven Press, 1992.

13. Galloway DA. Serological assays for the detection of HPV antibodies. In: Munoz N, Bosch FX, Shah KV, Meheus A, eds. The Epidemiology of Cervical Cancer and Human Papillomavirus. IARC Scientific Publications, Lyon: International Agency for Research on Cancer 1992; 119:147-161.

14. Gissmann L, Müller M. The current role of HPV-serology. In: Stanley M, Stern P, eds. Human Papillomaviruses and Cervical Cancer. Oxford: Oxford University Press, 1994:133-144.

15. Gissmann L. The current role of HPV serology. In: von Krogh G, Gross G, eds. Human Papillomavirus Infections in Dermatovenereology. Boca Raton: CRC Press, 1996:365-374.

16. Viscidi RP, Sun Y, Tsuzali B et al. Serologic response in human papillomavirus-associated invasive cervical cancer. Int J Cancer 1993; 55:780-784.

17. Sun Y, Eluf-Neto J, Bosch FX et al. Human papillomavirus-related serological markers of invasive cervical carcinoma in Brazil. Cancer Epidemiol Biomarkers Prevention 1994; 3:341-347.

18. Nindl I, Benitez-Bribiesca L, Berumen J et al. Antibodies against linear and conformational epitopes of the human papillomavirus (HPV)

type 16 E6 and E7 oncoproteins in sera of cervical cancer patients. Arch Virol 1994; 137:341-353.

19. Baay MF, Duk MP, Walboomers JMM et al. Antibodies to human papillomavirus type 16 E7 related to clinicopathological data in patients with cervical carcinoma. J Clin Pathol 1995; 48:410-414.

20. Baay MF, Duk JM, Burger MP et al. Follow-up of antibody responses to human papillomavirus type 16 E7 in patients treated for cervical carcinoma. J Med Virol 1995; 45:342-78.

21. Suchankova A, Ritterova L, Krcmar M et al. Comparison of ELISA and Western Blotting for HPV16 E7 antibody determination. J Gen Virol 1991; 72:2577-2581.

22. Köchel HG, Monazahian M, Sievert K et al. Occurrence of antibodies to L1, L2, E4 and E7 gene products of human papillomavirus types 6b, 16 and 18 among cervical cancer patients and controls. Int J Cancer 1991; 48:682-688.

23. Mandelson MT, Jenison SA, Sherman KJ et al. The association of human papillomavirus antibodies with cervical cancer risk. Cancer Epidemiol Biomarkers Prevention 1992; 1:281-286.

24. Kanda T, Onda T, Zanma S et al. Independent association of antibodies against human papillomavirus type 16 E1/E4 and E7 proteins with cervical cancer. Virology 1992; 190:724-732.

25. Onda T, Kanda T, Zanma S et al. Association of the antibodies against human papillomavirus 16 E4 and E7 proteins with cervical cancer positive for human papillomavirus DNA. Int J Cancer 1993; 54:624-628.

26. Paez CG, Yaegashi N, Sato S et al. Prevalence of serum IgG antibodies for the E7 and L2 proteins of human papillomavirus type 16 in cervical cancer patients and controls. Tohoku J Exp Med 1993; 170:113-121.

27. Mann VM, Loo de Lao S, Brenes M et al. Occurrence of IgA and IgG antibodies to select peptides representing human papillomavirus type 16 among cervical cancer cases and controls. Cancer Res 1990; 50:7815-7819.

28. Bleul C, Müller M, Frank R et al. Human papillomavirus type 18 E6 and E7 antibodies in human sera: Increased anti-E7 prevalence in cervical cancer patients. J Clin Microbiol 1991; 29:1579-1588.

29. Krchnak V, Pistek T, Vagner J et al. Identification of seroreactive epitopes of human papillomavirus type 18 E7 protein by synthetic peptides. Acta Virol 1993; 37:395-402.

30. Jha P, Beral V, Peto J et al. Antibodies to human papillomavirus and to other genital infections and their relation to invasive cervical cancer risk. The Lancet 1993; 341:1116-1118.

31. Dillner J, Lenner P, Lehtinen M et al. A population-based seroepidemiological study of cervical cancer. Cancer Res 1994; 54:134-141.

32. Sasagawa T, Inoue, M, Tanizawa O et al. Identification of antibodies against human papillomavirus type 16 E6 and E7 proteins in sera of patients with cervical neoplasia. Jap J Cancer Res 1992; 83:705-713.

33. Fisher SG, Benitez-Bribiesca L, Nindl I et al. The association of human papillomavirus type 16 E6 and E7 antibodies with stage of cervical cancer. Gynecol Oncol 1996; 61:73-78.

34. Eschle D, Dürst M, ter Meulen J et al. Geographic dependence of sequence variations in the E7 gene of human papillomavirus type 16. J Gen Virol 1992; 73:1829-1832.

35. ter Meulen J, Schweigler AC, Eberhardt HC et al. Sequence variation in the E7 gene of human papillomavirus type 18 in tumor and nontumor patients and antibody response to a conserved seroreactive epitope. Int J Cancer 1993; 53:257-259.

36. Wank R, Schendel DJ, Thomssen C. HLA antigens and cervical carcinoma. Nature 1992; 356:22-23.

37. Helland A, Borresen AL, Kristensen G et al. DQA1 and DQB1 genes in patients with squamous cell carcinoma of the cervix: relationship to human papillomavirus infection and prognosis. Cancer Epidemiol Biomarkers Prev 1994; 3:479-486.

38. Gregoire L, Lawrence WD, Kukuruga D et al. Association between HLA-DQB1 alleles and risk for cervical cancer in African-American women. Int J Cancer 1994; 57:504-537.

39. Apple RJ, Erlich HA, Klitz W et al. HLA DR-DQ associations with cervical carcinoma show papillomavirus-type specificity. Nat Genet 1994; 6:157-162.

40. Apple RJ, Becker TM, Wheeler CM et al. Comparison of human leukocyte antigen DR-DQ disease associations found with cervical dysplasia and invasive cervical carcinoma. J Natl Cancer Inst 1995; 87:427-436.

41. Nawa A, Nishiyama Y, Kobayashi T et al. Association of human leukocyte antigen-B1*03 with cervical cancer in Japanese women aged 35 years and younger. Cancer 1995; 75:518-521.

42. Jenison SA, Yu XP, Valentine JM et al. Evidence of prevalent genital-type human papillomavirus infections in adults and children. J Inf Dis 1990; 162:60-69.

43. Kaye JN, Starkey WG, Kell B et al. Human papillomavirus type 16 in infants: use of DNA sequence analyzes to determine the source of infection. J Gen Virol 1996; 77:1139-1143.

43a. Shah KV, Howley PH. Papillomaviruses. In: Fields BN et al. Virology. Philadelphia: Lippincott-Raven Publishers, 1996; 2:2077-2109.

44. Chambers MA, Stacey SN, Arrand JR et al. Delayed-type hypersensitivity response to human papillomaviruses type 16 E6 protein in a mouse model. J Gen Virol 1994; 75:165-169.

45. Höpfl RM, Sandbichler M, Sepp N et al. Skin test for HPV16 proteins in cervical intraepithelial neoplasia. Lancet 1991; 337:373-374.

46. Comerford SA, McCance DJ, Dougan G et al. Identification of T-and B-cell epitopes of the E7 protein of human papillomavirus type 16. J Virol 1991; 65:4681-4690.

47. Shepherd PS, Tran TTT, Rowe AJ et al. T cell responses to the human papillomavirus type 16 E7 protein in mice of different haplotypes. J Gen Virol 1992; 73:1269-1274.

48. Tindle RW, Fernando GJP, Sterling JC et al. A "public" T-helper epitope of the E7 transforming protein of human papillomavirus 16 provides cognate help for several B-cell epitopes from cervical cancer-associated human papillomavirus genotypes. Proc Natl Acad Sci USA 1991; 88:5887-5891.

49. Tindle RW, Herd K, Londono P et al. Chimeric Hepatitis B core antigen particles containing B- and Th-epitopes of human papillomavirus type 16 E7 protein induce specific antibody and T-helper responses in immunised mice. Virology 1994; 200:547-557.

50. Tindle RW, Croft S, Herd K et al. A vaccine conjugate of ëISCARí immunocarrier and peptide epitopes of the E7 cervical cancer associated protein of human papillomavirus type 16 elicits specific Th1- and Th2-type responses in immunized mice in the absence of oil-based adjuvants. Clin Exp Immunol 1995; 101:265-271.

51. Gao L, Chain B, Sinclair C, Crawford L et al. Immune response to human papillomavirus type 16 E6 in a live vaccinia vector. J Gen Virol 1994; 75:157-164.

52. Cubie HA, Norval M, Crawford L et al. Lymphoproliferative response to fusion proteins of human papillomavirus in patients with cervical intraepithelial neoplasia. Epidemiol Infect 1989; 103:625-632.

53. Kadish AS, Romney SL, Ledwidge R et al. Cell-mediated immune responses to E7 peptides of human papillomavirus (HPV) type 16 are dependent on the HPV type infecting the cervix whereas serological reactivity is not type-specific. J Gen Virol 1994; 75:2277-2282.

54. Altmann A, Jochmus-Kudielka I, Frank R et al. Definition of immunogenic determinants of the human papillomavirus type 16 nucleoprotein E7. Eur J Cancer 1992; 28:326-333.

55. de Gruijl TD, Bontkes HJ, Stukart MJ et al. T cell proliferative responses against human papillomavirus type 16 E7 oncoprotein are most prominent in cervical intraepithelial neoplasia patients with persistent viral infection. J Gen Virol 1996; 77:2183-2191.

56. Strang G, Hickling JK, McIndoe GA et al. Human T cell responses to human papillomavirus 16 L1 and E6 synthetic peptides: Identification of T cell determinants, HLA-DR restriction and virus-type specificity. J Gen Virol 1990; 71:423-431.

57. McLean CS, Sterling JS, Mowat J et al. Delayed-type hypersensitivity response to human papillomavirus type 16 E7 protein in a mouse model. J Gen Virol 1993; 74:239-245.

58. Chambers MA, Wej Z, Coleman N et al. "Natural" presentation of human papillomavirus type-16 E7 protein to immunocompetent mice results in antigen-specific sensitization or sustained unresponsiveness. Eur J Immunol 1994; 24:738-745.
59. Stauss HJ, Davies H, Sadovnikova E et al. Induction of cytotoxic T lymphocytes with peptides in vitro: Identification of candidate T-cell epitopes in human papilloma virus. Proc Natl Acad Sci USA 1992; 89:7871-7875.
60. Feltkamp MCW, Smits HL, Vierboom MPM et al. Vaccination with cytotoxic T lymphocyte epitope-containing peptide protects against a tumor induced by human papillomavirus type 16-transformed cells. Eur J Immunol 1993; 23:2242-2249.
61. Feltkamp MCW, Vreugdenhil GR, Vierboom MPM et al. Cytotoxic T lymphocytes raised against a subdominant epitope offered as a synthetic peptide eradicate human papillomavirus type 16-induced tumors. Eur J Immunol 1995; 25:2638-2642.
62. Sadovnikova E, Zhu X, Collins SM, Zhou J et al. Limitations of predictive motifs revealed by cytotoxic T lymphocyte epitope mapping of the human papilloma virus E7 protein. International Immunol 1994; 6:289-296.
63. Ossevoort MA, Feltkamp MCW, van Veen KJH et al. Dendritic cells as carriers for a catotoxic T-lymphocyte epitope-based peptide vaccine in protection against a human papillomavirus type 16-induced tumor. J Immunother 1995; 18:86-94.
64. Zhu X, Tommasino M, Vousden K et al. Both immunization with protein and recombinant vaccinia virus can stimulate CTL specific

70. Kast WM, Brandt RMP, Sidney J et al. Role of HLA-A motifs in identification of potential CTL epitopes in human papillomavirus type 16 E6 and E7 proteins. J Immunol 1994; 152:3904-3912.
71. Ressing ME, Sette A, Brandt RMP et al. Human CTL epitopes encoded by human papillomavirus type 16 E6 and E7 identified through in vivo and in vitro immunogenicity studies of HLA-A*0201-binding peptides. J Immunol 1995; 154:5934-5943.
71a. Jochmus I, Osen W, Altmann A, Buck G, Hofmann B, Schneider A, Gissmann L, Rammensee H-G. Specificity of human cytotoxic T lymphocytes induced by a human papillomavirus type A6 E7-derived peptide. J Gen Virol 1997; 78:in press.
72. Ressing ME, van Driel WJ, Celis E et al. Occasional memory cytotoxic T-cell responses of patients with human papillomavirus type 16-positive cervical lesions against a human leukocyte antigen - A*0201-restricted E7-encoded epitope. Cancer Res 1996; 56:582-588.
73. Bartholomew JS, Stacey SN, Coles B et al. Identification of a naturally processed HLA A0201-restricted viral peptide from cells expressing human papillomavirus type 16 E6 oncoprotein. Eur J Immunol 1994; 24:3175-3179.
74. Tarpey I, Stacey S, Hickling J et al. Human cytotoxic T lymphocytes stimulated by endogenously processed human papillomavirus type 11 E7 recignize a peptide containing a HLA-A2 (A*0201) motif. Immunology 1994; 81:222-227.
75. Gissmann L. Human papillomaviruses and genital cancer. Semin Cancer Biol 1992; 3/5:253-261.
76. Cromme FV, Meijer CJLM, Snijders PJF et al. Analysis of MHC class I and II expression in relation to presence of HPV genotypes in premalignant and malignant cervical lesions. Br J Cancer 1993; 67:1372-1380.
77. Wu T-C, Guarnieri FG, Staveley-OíCarroll KF et al. Engineering an intracellular pathway for major histocompatibility complex class II presentation of antigens. Proc Natl Acad Sci USA 1995; 92: 11671-11675.
78. Lin K-Y, Guarnieri FG, Staveley-OíCarroll KF et al. Treatment of established tumors with a novel vaccine that enhances major histocompatibility class II presentation of tumor antigen. Cancer Res 1996; 56:21-26.
79. Boursnell MEG, Rutherford E, Hickling JK et al. Construction and characterization of a recombinant vaccinia virus expressing human papillomavirus proteins for immunotherapy of cervical cancer. Vaccine 1996; 14:1485-1494.
80. Borysiewicz LK, Fiander A, Nimako M et al. A recombinant vaccinia virus encoding human papillomavirus types 16 and 18, E6 and E7 proteins as immunotherapy for cervical cancer. Lancet 1996; 347:1523-1527.

81. Tindle RW. Human papillomavirus vaccines for cervical cancer. Curr Opin Immunol 1996; 8:643-650.
82. Thomson SA, Khanna R, Gardner J et al. Minimal epitopes expressed in a recombinant polyepitope protein are processed and presented to CD8+ cytotoxic T-cells. Proc Natl Acad Sci USA 1995; 13:5845-5849.
83. Pozzi G, Contorni M, Oggioni MR et al. Delivery and expression of a heterologous antigen on the surface of streptococci. Infect Immunity 1992; 60:1902-1907.
84. Oggioni MR, Manganelli R, Contorni M et al. Immunization of mice by oral colonization with live recombinant commensal streptococci. Vaccine 1995; 13:775-779.

INDEX

A

Activation domain, 36, 38, 122
Activator, 30, 38, 42, 45
  AP1, 43, 58
  of P97, 49
  pol I, 121
  proteins, 43
  transcriptional, 36, 82
Adenovirus (Ad), 13, 39, 48, 76-78,
  86-87, 121, 158
  5, 120
  E1A, 104, 120, 122, 125
  E2, 38, 122
Anogenital
  cancer, 10, 78, 156
  intraepithelial
    neoplasia, 7
    cancer, 7
  lesions, 72
  specific HPVs, 73
  tract, 9, 15, 103
  types, 6
Antibody, 141, 143
  anti-E7 polyclonal, 110
  antigen
    complex, 139, 141-142
    reaction, 139
  detection, 146
  increased prevalence, 145
  monoclonal, 141, 143
  response, 137
  titers, 146
Antigen presenting cells (APC), 137-138,
  148, 157
Antisense, 92
  constructs, 91
  E6/E7 constructs, 12
  oligonucleotides, 92
  sequences, 92
AP1, 35, 39-43, 45-47, 58-59, 114, 121
  binding site. *See* Binding sites.
Apoptosis, 74-76, 85, 88, 125
  E7-induced, 86, 88, 124
  in cells devoid of E1B, 88
  of HPV-infected keratinocytes, 87
  p53-mediated, 88-90
  reduction, 86

ATF, 39, 121
5-Azacytidine, 56

B

β-galactosidase, 110
B-myb, 120
Basal
  cell, 8-9, 25, 27, 47, 58, 87
  transcription apparatus, 29-30
Basic zipper motif, 43
Bax, 74-75
bcl-2, 75
BHK-21 cell, 110
Binding sites, 35, 41, 44, 47
  AP1, 41, 43, 45
  C/EBP, 43
  DNA, 30, 45
  E2, 31, 35-37, 44
  E2F, 39, 120
  Epoc-1, 47
  for cellular transcription factors, 31,
    35
  NF1, 43, 48
  Oct-1, 46, 47
  promoter-proximal YY1, 42
  Sp1, 45
  TEF-1, 49
  transcription factor, 35, 37, 40
  YY1, 35, 42, 43
Bovine papillomavirus (BPV), 4, 36-38,
  54, 84-85

C

c-fos, 59, 122
c-fos, 40
c-myc, 112
Cancers
  esophageal, 15
  oropharyngeal, 15
  of the respiratory route, 15
Capture ELISA, 141
Casein kinase II (CKII), 105, 108, 115
CaSki cells, 53, 103, 111, 155, 157. *See
  also* Cervical carcinoma.
CCAAT/enhancer binding protein, 42-43

Cell
    cycle, 14, 30, 38-39, 56, 75-76, 83,
        87-88, 90, 113-117, 119, 124
    hybridization, 12, 55, 57
Cell-mediated immunity, 138
Cervical carcinoma, 7, 12, 27-29, 42, 46,
    48, 50, 52-53, 55-57, 59, 71, 103,
    109-110, 113, 155, 157
    cells, 109
        CaSki, 110. *See also* CaSki cells.
        HeLa, 52, 103. *See also* HeLa cells.
        SiHa, 103, 110. *See also* SiHa cells.
Cervical intraepithelial neoplasia (CIN),
    10, 15, 27-28, 55, 58, 151, 155-156
Chimeric transcript, 50
Chlamydia trachomatis, 145
Chromatin, 30, 52-53
Chromosome 11, 55-56, 59
Cisplatin, 84-85, 119
Coactivator, 30, 38, 42, 49
Composite regulatory element, 47, 49
Condylomata acuminata, 5
Conserved regions, 105
    1, 104
    2, 104
    3, 104
Constitutive enhancer. *See* Enhancer.
COS-1, 109-110
Cottontail rabbit papillomavirus
    (CRPV), 3, 54
^{51}Cr release assay, 148-149
CXXC motif, 73, 108-109, 123
Cyclic AMP, 122
Cyclin, 79, 117, 119
    A, 14, 114, 116, 119-120, 123
    D, 56, 116, 119, 123. *See also* D type
        cyclin.
    D1, 120, 123
    dependent kinase (CDK), 74, 88, 115,
        116
        CDK4, 116, 123
        CDK6, 116
    dependent kinase inhibitor, 56
    dependent kinases, 75, 79
    E, 14, 114, 116, 119-120, 123
    gene expression, 120

D

D type cyclin, 116. *See also* Cyclin.
Degradation
    CDK inhibitors, 116

cytosolic and nuclear proteins, 79
    p53, 82, 86, 92, 124
        E6-induced, 13, 77-82
        in vitro, 86
    pRb, 120
    selective, 79
Delayed type hypersensitivity (DTH),
    148, 152
Dendritic cell, 137, 153, 158
Dexamethasone, 45, 53, 113
Diagnosis
    early, 14
    of HPV associated diseases, 152
    serology for clinical, 144
Differentiation, 14, 25, 27, 35, 38, 41, 43,
    45-46, 48, 56, 73, 87, 89, 119, 124, 126
Dimerization, 36, 43, 105, 109
DNA
    binding domain, 48
    damage, 75, 79, 83, 88, 92, 117, 120
    methylation, 53-54
    replication, 4, 25, 27, 35, 39-40, 43-44,
        48, 89-90, 113, 115
    binding domain, 37
DNAse I hypersensitive site, 52
DP gene, 116-117, 119-120

E

E1, 30, 35-36, 50, 71-72, 76, 79-81
    A, 88, 104-105, 120, 122, 125
    B, 78, 88
E2, 26, 31, 36-38, 44, 50, 79-80, 84, 90-91
    F, 13, 39, 116, 120-121
    F-1, 116, 119-120
    F/DP complexes, 117
    F/DP heterodimer, 119
    gene of adenovirus 5, 120
    open reading frame (ORF), 11, 50, 71
    promoter, 38
E3, 79, 80
E4, 31, 78, 144, 147
E5, 31, 112, 125
E6, 11, 36, 83
    amino acids, 153
    antibodies, 144-147
    antigens, 158
    antisense constructs, 12
    as potential therapeutic vaccine,
        156-165
    associated protein (E6-AP), 78, 81-82,
        87, 92. *See also* E6-AP.

binding protein (E6BP), 87
cellular immune responses against,
    147-156
cooperation with E7, 89
core promoter, 31
correlation of antibodies of with
    cervical cancer, 144-146
deregulation, 26
downregulation, 58
elevated RNA levels, 27
expression, 25, 27, 43, 50, 52-54,
    56-57, 73, 84-85, 89-91, 113
  regulation, 48
functional consequences of interaction
    with p53, 82-102
high risk, 73, 77, 82-83, 85-89, 92, 113
HPV16, 86, 88, 124, 145-146, 151-155,
    157
HPV18, 83-84, 151, 157
humoral immune response against,
    138-147
in vivo expression, 26-27
independent pathways, 86
induced degradation of p53, 77-82
induced ubiquitination of p53, 80
inhibition of expression, 26, 71
interplay with E7, 87-102
location, 30-31
low risk, 74, 82-83, 86-87
open reading frame (ORF), 71-73
p53-independent properties of, 85-87
promoter, 31, 36-37, 39-42, 44-45,
    47-48, 90-91
protein, 38-39, 52, 72-74, 85, 90,
    112-113, 125, 138, 146, 154
recombinant vaccinia virus, 154, 157
regulation of transcription, 25-26, 36,
    38, 44
role in
    degradation of p53, 13
    viral life cycle, 89-90
suppression of transcription, 41
therapeutic responses, 90-102
transcription, 26, 48, 58, 84, 90
transcripts, 27-29, 92
transgene, 88
upregulation, 27
viral and cellular factors for transcrip-
    tion, 29
viral DNA integration for, 26
E6*, 28, 72
E6-AP, 78, 81-82, 87. See also E6.

E7, 11, 36, 39, 55, 90
antisense constructs, 12
deregulation of expression, 26
expression, 25, 27, 37, 43, 50, 52-54,
    56-58, 84-85, 89-90
functional interplay with E6, 87-102
high risk, 88
HPV16, 29, 84, 86
    transcripts, 28
HPV18, 84
in vivo expression, 26-28
induced apoptosis, 86, 88
inhibition of expression, 26, 71
location, 30
open reading frame (ORF), 50, 71-72
promoter, 90
protein, 13, 39-40, 52, 73, 77, 88, 90
regulation of
    expression, 48
    transcription, 25-26, 36, 38, 44
RNA levels, 27
suppression of transcription, 41
therapeutic perspectives, 90-102
transcription, 48, 58, 90
transcripts, 27-28
upregulation, 27
viral and cellular factors for transcrip-
    tion, 29-49
viral DNA integration for, 26
Early and late viral gene expression, 26
Early regions, 30
ELISA, 138, 140-142, 144-145
capture. See Capture ELISA.
Enhancer
    -promoter constructs, 38
    activity, 35-36
    autonomous, 36
    CCAAT, 42
    central URR, 36
    constitutive, 35, 40
    E2-responsive, 36
    E6-responsive, 38
    element(s), 30-31, 36
Epidermodysplasia verruciformis (EV),
    6-8, 11, 72
Epithelial cells, 35-36, 43-44, 46, 48-49,
    85-86, 112, 117, 125, 137, 156
Epoc-1, 47-48
Epstein-Barr virus, 14, 77-78
ERC-55, 87
Established rodent fibroblast lines, 111

F

Folic acid, 54
fos, 40-41, 112
Fra-1, 59
Freund's adjuvant, 153, 158
Fusion transcript, 51

G

G1 arrest, 75, 117, 119-120
Gastric cancer, 14
General transcription initiation factors, 29
    TFIID, 30, 37
Genetic instability, 83
Glucocorticoid
    receptors, 45
    regulated response, 52
    responsive element (GRE), 45

H

H2-Kb, 153-154
Half-life, 77
    fusion transcript, 51
    p53, 77, 85-86
    virus-specific mRNA, 50
Hect domain, 82
HeLa cells, 42-43, 46, 53, 55-56, 58, 84, 87. *See also* Cervical carcinoma.
Helicobacter pylori, 14
Hepatitis B virus, 77-78
Heterodimer, 37, 41, 45, 59, 116, 119-120
HLA, 147, 152, 155, 158
Human papillomavirus (HPV), 3-5, 25, 54, 89 123
    HPV DNA replication, 27, 48
    HPV type(s), 5, 11, 16, 104
    HPV4, 108-109
    HPV5, 85, 106, 111-112
    HPV6, 27, 85-86, 106, 113, 122, 147
    HPV7, 106
    HPV8, 85, 106, 111, 112
    HPV11, 27, 43, 45-46, 83, 122, 147, 156
    HPV13, 106
    HPV16, 9-10, 12, 105
    HPV18, 27, 31, 35-38, 40-42, 44-47, 49-50, 52-54, 56-57, 82, 84, 119, 144-146, 157
    HPV24, 109

HPV29, 106
HPV40, 106
HPV41, 108-109
HPV44, 106
HPV48, 108
HPV49, 109
HPV54, 109
HPV55, 106
HPV74, 106
Humoral immune response, 138, 143-144, 146
Hypermethylation, 56
Hypomethylation, 54

I

Immortalization, 12, 37, 50, 71, 73-74, 87, 103, 106, 112-113, 123

J

jun, 40-41

K

Koilocytotic atypia, 10

L

L1 protein, 3
Late gene, 25, 27, 36
    expression, 27
Late regions, 31
Latency period, 11, 55
Long control region, 31
LXCXE domain, 106
Lymphoma
    B-cell, 14, 78
    T-cell, 78

M

Macrophage-chemoattractant protein-1 (MCP-1), 13
Major histocompatibility complex (MHC)
    class I complexes, 156
    class I molecules, 156
    class II alleles, 150
    class II molecules, 148

class II processing and presentation, 157
class II restricted T cells, 157
transport proteins, 13
Mitomycin C, 84, 119

N

Nonanogenital cancer, 15
Noncoding regions, 31
Nonmelanoma skin cancer. *See* Skin cancer.
Nonselective conditions, 57
Nuclear factor I (NFI), 35, 43, 47-49
  NF-IL6, 43, 58
Nucleoli
  of CaSki cells, 111
Nude mice, 54, 111, 113, 154

O

Oct protein, 46, 48
Oct-1, 35, 39, 46-47, 49, 58, 121
Octamer-binding, 46
Okadaic acid, 56
Origin of replication, 31

P

p15INK4B, 116, 119
p16, 56
p16INK4, 116, 123
p16INK4a, 116, 119
p21WAF1, 74-75, 88-89, 116, 119, 123
p27KIP1, 115-116, 119, 123
p53, 75, 79
  activation, 75, 79
  as target of DNA tumor viruses, 76-77
  consequences of interaction with E6, 82-102
  degradation, 13, 83, 86, 125
    E6 induced, 77-82
  expression, 74
  function, 38, 76, 83-84, 117
  gene, 83, 112
    mutation, 76
  half-life, 85
  in vivo, 79
  inactivated, 76
  inactivation, 77
  independent properties of E6, 85-87

induced apoptosis, 75
induced G1 arrest, 75
mediated apoptosis, 88-89
mediated genomic stabilization, 89
mutant, 76, 117
mutants, 83, 85
mutation of locus, 74
null cells, 76
null mice, 76, 124
overexpression, 84
pathways, 84, 124
protein, 13, 38, 76-77, 82, 84
transcriptional repression, 75
ubiquitination of, 80
virus interaction with, 78
wild-type, 74
p53 and HPV status, 84
p53-mediated transcriptional regulation, 82, 83
P_{97}, 31, 35, 38, 43, 49, 72
P_{105}, 31, 35, 42-45, 72
p107, 39, 106, 108, 114-116, 119-120, 123
p130, 106, 108, 114-116, 119
p300, 104
Papillomavirus enhancer binding factor 1 (PEF-1), 47
Paracrine regulation, 12
Peptide(s), 141-142, 145, 148, 150-152, 156
  9-mer, 153
  10-mer, 153-154
  20-mer, 150
  antigens, 145
  chemically synthesized, 138
  chosen for vaccination, 153
  derived from E6-AP, 92
  E6-derived, 150
  E7, 157
  E7 CR2, 121
  E7-derived, 106, 155
  ELISA, 144, 146
  externally loaded, 149
  fractions, 154
  HPV16 E6-derived, 152, 156
  HPV16 E7-derived, 151, 158
  HPV16 E7-specific, 145
  HPV18 E7, 145
  HPV18 E7-derived, 145
  HPV18-derived, 145
  microinjection studies, 122
  nonamer, 155

synthetic, 141, 146, 155-156, 158
vaccination, 158
Peripheral blood lymphocytes (PBL), 147
Persistent infection, 25, 144, 151
Polyadenylation signal, 52
pRB, 77
  independent pathways, 39
  protein, 13, 39
pRb (tumor suppressor gene
    retinoblastoma), 88-91, 104, 106, 108,
    110-112, 114-116, 119-122
  binding domains, 107, 124
  E7 interaction, 107
Primary human keratinocytes, 50, 54,
    112, 117
Primary rodent cells, 106, 112
Progesterone receptors, 45
Promoters, 29-31, 40, 120
  adenovirus E2, 122
  BPV1 E6, 36
  c-fos, 122
  cellular, 52
  core, 29
  E2, 38, 122
  E2F-responsive, 39
  E6, 31, 36-37, 39-41, 44-46, 48
  E6 core, 31
  E6/E7, 91
  early, 46
  early gene, 39
  enhancer-, 38
  heterologous, 38, 40, 45, 49, 83, 86,
    120
    systems, 112
  HPV E6, 37
  HPV16 P_{97}, 38
  HPV18 E6, 42
  HPV18 E6/E7, 90
  interstitial retinol-binding protein,
    124
  murine aA crystalline, 124
  P_{89}, 36
  P_{97}, 43, 49
  P_{105}, 35, 42-45
  proximal sites, 44
    Sp1 site, 44
    YY1 binding site, 42
  regions, 30, 43-44
  target, 86
  TATA-containing, 38
  URR E6, 40
  viral, 28

Protein kinase C (PKC), 41
Protein phosphatase 2A (PP2A), 12, 56
PVF, 49

R

Rabbit papillomavirus, 2-3, 54
Radio-immunoprecipitation (RIPA),
    142-143, 146
ras gene product, 112
Rat 3Y1 cells, 110
Recombinant vaccinia virus, 110, 150,
    152-154, 156-157
Recombination, 50, 52-53
Regions. *See* Conserved regions, Early
    regions, Late regions, Noncoding
    regions.
Repression, 12, 30, 37, 42-43, 74, 75, 90,
    122
Repressor, 26, 30, 37-38, 42-43, 56, 82, 91
Retinoblastoma, 77, 104, 116, 119-120
Retinoic acid receptor, 58
Retinoid receptor, 44-45
Ribozyme, 91
RNA
  labilization signal, 51
  polymerase II, 29-30, 121
  -RNA in situ hybridization, 26-29

S

Schizosaccharomyces pombe, 110
Secondary antibodies, 139-140
Serology, 143-144
Serum, 73, 139, 141
  growth factors, 117
  standard, 140
  starved cells, 120
  withdrawal, 119
Shope papillomavirus, 11
SiHa cells, 52, 55, 57, 111. *See also*
    Cervical carcinoma.
Simian virus 40 (SV40), 13, 49, 76-78
  immortalized cells, 12
  large TAg, 125
  TAg (tumor antigen), 76-77, 104
  T antigen expression, 12
Skin cancer
  associated with EV, 85
  epidermodysplasia verruciformis, 6
  HPV types in, 8

model for infectious origin, 3
nonmelanoma, 8, 15
Skin types, 6
Somatic cell
hybrid, 56
hybridization, 12, 55
Squamous cell carcinoma, 3, 6-7, 9-10
Steroid hormone, 45
receptor, 35, 44-45
Streptococcus gordonii, 158
Subfractionation, 154
Switch region, 42-43

T

T helper memory, 151
T lymphocytes, 147-148, 150
cytotoxic, 148, 154-155
HPV16 E6-specific, 154
proliferation, 150-152
TAF110, 121
TATA
-binding protein, 30
box, 29, 31, 35, 37, 44, 52, 91, 108, 121
box binding protein (TBP), 108
-associated factor (TAF), 30, 39, 121
Telomerase, 86-87
Therapeutic vaccine, 138, 156, 158
Tolerance
HPV16 E7, 147
T-cell, 152
Transactivation domain, 37
Transformation, 10, 39, 85, 87-88,
103-104, 106, 108-109, 125
assays, 74, 106, 108
cell, 4, 45, 83, 87, 104, 106, 111-112,
123
cellular, 51, 72, 76-77, 86-87, 89
malignant, 11, 51, 103, 112
of infected cells, 71
of rodent cell lines, 111-112
to cervical cancer, 28
viral, 50
Transformation system, 74
Tumor
antigen. *See* Simian virus 40.
necrosis factor (TNF), 12

U

Ubiquitin system, 77, 79, 81

Ubiquitination, 80
E6-AP-mediated, 81
p53
E6-induced, 80-82
multi, 81
proteins, 79
Upstream
binding factor (UBF), 121
regulatory region (URR), 30-31,
35-50, 52-53, 56-57, 112
Uroepithelial cell, 83, 88, 124
UV irradiation, 119

V

Verrucosis, 6
Viral DNA replication, 25, 27, 35, 40, 90
Virion assembly, 25
Virus-like particle, 3, 14

W

Warts, 5, 138
bovine, 1, 4
Butchers', 7
canine, 1
common, 5, 7
confluent flat, 6
flat, 6-7
genital, 2, 5, 7, 10, 145
histological types of, 5
human, 2-3
in dogs, 2
infectious etiology of, 1
large horny, 2
myrmecia, 7
negative, 5, 10
nonmalignant, 6
pigmented, 7
plantar, 5, 7
viral etiology of, 2
Western blotting, 138-140

Y

YY1, 31, 35, 42-43, 49, 58-59

Z

Zinc fingers, 38, 44, 109, 123